MILLIKAN'S
SCHOOL

MILLIKAN'S SCHOOL

◆

A History of
The California Institute of Technology

JUDITH R. GOODSTEIN

W · W · Norton & Company
New York · London

Portions of this work appeared in different form in the journals *Science* and *California History*.

The chapter on earthquakes appeared originally in slightly different form in *Historical Studies in the Physical Sciences*.

The chapter on Linus Pauling appeared originally in slightly different form in *Social Research*.

Printed in the United States of America.

The text of this book is composed in Galliard
with the display set in Florentine.
Composition by PennSet.
Manufacturing by Courier Companies Inc.
Book design by Jack Meserole.

First Edition.

Library of Congress Cataloging-in-Publication Data
Goodstein, Judith R.
Millikan's school : a history of the California Institute of
Technology / Judith R. Goodstein ; foreword by Lee A. DuBridge (etc.)
p. cm.
Includes index and bibliography.
1. California Institute of Technology—History.   I. Title.
T171.C219
[G67 1991]
607.1′179493—DC20   91-6748

ISBN 0-393-03017-2

W.W. Norton & Company, Inc., 500 Fifth Avenue, New York, N.Y. 10110
W.W. Norton & Company, Ltd., 10 Coptic Street, London WC1A 1PU

1 2 3 4 5 6 7 8 9 0

For

DAVID

and

the memory of

SIG KORAL

(1916–1990)

CONTENTS

Photographs appear following pages 96 and 224.

PRESIDENTIAL PERSPECTIVES

♦

Institutions that achieve distinction do so by following certain basic principles. Institutions that continue to achieve distinction over long periods of time have to adapt, to change in ways that do not compromise the institution. Although Caltech has remained devoted to excellence, has remained small and focused on certain areas where it could make a difference, and has continued to be blessed with friends locally and across the nation, it has changed a great deal since Dr. Millikan's time. For example, Dr. Millikan was suspicious of federal support, and the Institute had very little while he led it. Under Dr. DuBridge, his worthy successor, the Institute's research received generous funding from the federal government, so that from the 1950s to the present day, approximately one-half of the campus funding has come directly as federal research support.

Dr. Goodstein's account of "Millikan's School" selects certain key individuals who helped make the Institute famous. They are well chosen. And yet, any such attempt at selection necessarily omits other contributions that were also key to the overall success of the enterprise. In particular, since faculty spend a good deal of their life at an institute such as Caltech, their integrated contributions are greater than those of the students. However, when one tries to assess the Institute's impact on the world, it becomes evident that the excellent grounding that Caltech students received and their resulting contributions surely had a leavening effect on the overall development of science and technology in many places. For example, since the early 1920s, Caltech has graduated over 1 percent of the Ph.D.'s in the United States in those fields in which it is active. Many of these individuals have become leaders in other institutions and have had a major impact on science and technology in their own right.

As Caltech begins its second century, it continues to believe that extraordinary advances in knowledge are made by extraordinary people, that ideas from one field can stimulate advances in another, that science and technology are mutually supportive, that new science gives rise to new technology, and that new technology, instrumentation, and computational power speed the progress of science. Opportunities seem virtually boundless in science and technology, and nowhere is this truer than at "Millikan's School."

THOMAS E. EVERHART
President
Pasadena, California
January 13, 1991

◆

The year 1891 does not appeal to most people as one of the important years in the history of science or education. And yet it was in 1891 in two different locations that two institutions were formed that were to play a large role in the development of science and education in the United States. Both of these institutions started out as manual training schools based upon the theory that one learns by doing, and both in their own ways were successful in their own communities. However, they were destined for better things.

John D. Rockefeller picked the city of Chicago as a likely place to establish a great university. He started with the manual training school as its first nucleus, and it developed, as everyone knows, into the University of Chicago, one of the country's great institutions. In Pasadena, the trade school was located in the second floor of a warehouse and was known as Throop University for its first year or two. Then it moved to its own campus a few blocks away and changed its name to Throop Polytechnic Institute. This was not its final step, however, because it soon became the Throop College of Technology, with a new location in a suburban part of the city of Pasadena. Throop College of Technology was a pioneering educational institution in the field of engineering in Pasadena.

Meanwhile, the Mount Wilson Observatory was established, under the leadership of George Ellery Hale, who had persuaded the Carnegie Institution of Washington to provide the funds for a new observatory. He saw that a new observatory, to be great, would need to have a college or university nearby. As Hale developed the Mount Wilson Observatory, he also became a member of Throop's board of trustees and persuaded the board to convert that institution into a major center of science and education. To accomplish this, it was necessary to bring to the institution an outstanding leader who would attract many other outstanding men in the fields of science and education. The man he chose was an old friend of his from their University of Chicago days, Robert A. Millikan. Millikan brought with him Earnest Watson, Ira Bowen, and others, all of whom played important roles in establishing the various fields of activity of the Institute.

Clearly, at this stage in the game, Throop College of Technology was no longer an appropriate name for an institution that hoped to have a much wider scope and that in the mind of Hale would be very much like his own alma mater, the Massachusetts Institute of Technology. Hence, the name was changed to the California Institute of Technology, now shortened to Caltech. Millikan himself did not particularly like the name, but following the arrival of Millikan and the coterie of outstanding people who were attracted there, Caltech grew in stature and reputation with astonishing rapidity over the period of the next thirty years or more.

One of the outstanding events of this time was the decision of the people at the Mount Wilson Observatory, under Hale, to build a very much larger telescope than the 100-inch giant at Mount Wilson. Hale proposed then to build a 200-inch telescope, but quickly realized that Pasadena was no longer a suitable location for such a large instrument, because growing traffic, growing smoke, and growing haze made seeing more and more difficult. The Rockefeller Foundation, under the leadership of Max Mason, a former professor of physics at the University of Wisconsin, and after that president of the University of Chicago, granted $6 million to the California Institute of Technology, to build a 200-inch telescope to be located on Palomar Mountain. The Palomar Observatory, completed and dedicated in 1949, was for forty years or more the outstanding astronomical instrument of the world.

This book is about the development of the California Institute of Technology, its scientific and engineering activities, its ventures into various fields of science and the humanities, and the out-standing group of professors, graduate students, and research fellows who became a part of the Caltech community. It is especially the story of the great men who made it possible for this institution to grow, to thrive, and to make such extraordinary contributions to the history of science and education. An institution is often said to be the shadow of a man. Caltech is the shadow of many men and women, but above all it is a tribute to Robert A. Millikan, who more than any other single person forged Caltech into the great scientific university that it is. In a very real sense, Caltech is "Millikan's School."

LEE A. DUBRIDGE
President Emeritus
Pasadena, California
December 5, 1990

◆

Caltech's is a fascinating story, and Judy Goodstein, as Institute archivist, knows where the papers and the tapes (if not the bodies) are buried. The history of Caltech's early days, from 1891 when it began its institutional life as Throop University, through Robert Millikan's time, and into Lee DuBridge's tenure displays some notable continuities. Already a few years before 1920, when Throop became Caltech, several themes were established that had resonance through the late 1960s and 1970s, and even into the 1990s: relations between the Institute and the federal government, and those between private and state-supported institutions of high learning; the tensions, interactions, and mutual support among trustees, faculty, and students; the continual search for sources of funds to begin new activities and to carry on existing ones—a search very satisfying when successful and very frustrating when less so; military-related research and development; questions of gender and of ethnicity; the need for scientists and engineers to be thoroughly exposed to the humanities in the Institute's curriculum. Perhaps two continuing threads are the strongest of all. The first one is Caltech's concentration and specialization in what have been seen as the most important fields at the frontiers of science and technology. The second is its emphasis on recruiting the best people to lead them; this has served as a criterion, along with the judgment of what would be important in the future, in deciding whether and when to move into new areas of teaching and research. Practically all of its general areas were established at Caltech relatively early: chemistry and physics, engineering and aeronautics, geology and biology.

A small number of leading figures set the style and the traditions that, modified by time and by circumstance, characterize Caltech to this day. The Institute was to be, and still is, small in numbers, gigantic in quality and in influence. It has reared—in its students, postdoctoral fellows, and junior faculty—leaders of other institutions in the United States and abroad. Moreover, Caltech has created larger activities out of itself—such as those of JPL—that apply technology on a grand scale.

The founders had their prejudices and personal quirks—possibly even more of the latter than appear in *Millikan's School*. But devotion to the goal of being at the forefront both of science— which means understanding nature—and of technology—controlling nature—has characterized Caltech from its beginning. That dedication is what physicists call "a constant of the motion" to this day and will remain so, we can be confident, through Caltech's second hundred years.

HAROLD BROWN
President Emeritus
Washington, D.C.
February 1, 1991

◆

There is no place like Caltech. The institution that has evolved from the heroic efforts of Millikan, Noyes, and Hale occupies a unique position in the research and educational establishment. If a single factor can be called responsible for this, it is Caltech's absolute unwillingness to compromise on excellence: excellence of faculty, excellence of students, excellence of facilities.

Everyone at Caltech is proud to be associated with it. To be accepted as a student, you need near-perfect SAT scores, so you are immediately among the elite. Invitations to join the faculty come only after the most excruciating examination. There are no good appointments at Caltech—only superb ones. Of course, occasional mistakes occur, but overall the level of quality is amazingly high, and, as many prestigious institutions have learned, it is very difficult to lure people away. Caltech has always recognized the importance of excellent facilities and equipment. The Mount Wilson Observatory, the early Van de Graaff accelerator, the wind tunnels, the Mount Palomar Observatory, the Owens Valley Radio Observatory, the Big Bear Solar Observatory, the synchrotron laboratory, and more recently the Keck Observatory and the gravitational radiation detection program are examples of bold initiatives and commitments of resources unmatched by any other institutions. All of the people who maintain these facilities, keep the campus sparkling, and administer the myriad services to faculty and students share in the exhilaration of being associated with the very best.

One of Caltech's greatest assets is the Jet Propulsion Laboratory. Begun in the latter part of Millikan's reign, this integral part of Caltech has been of immense importance both intellectually and financially. It is the world center for Solar System exploration, and the spectacular pictures of the *Voyager* spacecraft's journey have been seen in every part of the globe. Managing this huge enterprise is a serious responsibility which Caltech has done very well.

Aside, perhaps, from the heavy dependence on federal funding, Millikan would probably be pleased with what his child has grown into. By being bold, imaginative, and dedicated to excellence, even though small, Caltech casts a giant shadow.

MARVIN L. GOLDBERGER
President Emeritus
Princeton, New Jersey
January 30, 1991

ACKNOWLEDGMENTS

♦

There is a great mystery surrounding institutions of higher learning: why do some schools succeed and flower, while others remain in place, go nowhere, even wither away?

In the case of Caltech, it was the inspired leadership of Hale, Millikan, and Noyes that brought the school to the pinnacle of excellence. They formed a potent combination, these three men, working together to build Caltech, for they were neither hustlers nor bench scientists. Hustlers can raise money; Millikan, Noyes, and Hale did more than just raise money. Great scientists themselves, they could also pick scientists and inspire their students to do great science. Prodigious correspondents, they left behind a rich legacy of personal and scientific letters, the starting point for this book.

In writing this history, I owe much to Dr. John Greenberg, who participated in the research, gathered material, interviewed and corresponded with scientists, and put his deep knowledge of the history of science at my disposal. Research for the book was supported by a grant (1980–84) from the John Randolph Haynes and Dora Haynes Foundation, and it is a pleasure to thank them for their generous support of my project. An additional research grant (1982) from the Rockefeller Archive Center of the Rockefeller University greatly facilitated my work in the center's manuscript collections.

For permission to quote from unpublished materials, I thank the following: Beth Carroll-Horrocks, American Philosophical Society; Robin Rider, Bancroft Library, William M. Roberts, University Archives, and John Heilbron, Office for History of Science and Technology, UC Berkeley; Harry Ransom Humanities Research Center, University of Texas at Austin; the Alan Mason Chesney Medical Archives of the Johns Hopkins Medical

Institutions; Darwin H. Stapleton, Rockefeller Archive Center; Daniel Meyer, University of Chicago Library; Richard T. Black, Carnegie Institution of Washington; Judith Ann Schiff, Yale University Library; Nancy L. Boothe, Rice University Archives, Woodson Research Center, Rice University Library; Ze'ev Rosenkranz, Albert Einstein Archives, Hebrew University of Jerusalem; and Spencer Weart, American Institute of Physics. Grateful acknowledgment is made to Robert F. Bacher for permission to quote from his oral history memoir and to Lee A. DuBridge to quote from his unpublished recollections, both in the Caltech Archives.

Janet Jenks, Dana Roth, and Carolyn Garner answered all my reference questions. Jeanne Tatro and Tess Legaspi tracked down books at libraries around the country. Johanna Tallman and Glenn Brudvig didn't ask too many questions. Carol Bugé dug through old ledgers and letter files. Paula Hurwitz cared for other scholars. Gwen Torges organized the book's illustrations. Loma Karklins transcribed the oral histories. Bonnie Ludt pored over the footnotes. Heidi Aspaturian's eagle eye and deft pen salvaged more than one impenetrable paragraph. Stanley Goldberg introduced me to the National Archives and scouted the Merle Tuve papers. Michael Meo, Albert C. Lewis, Thomas Rosenbaum, Sharon Gibbs Thibodeau, Marjorie Ciarlante, and Joan Warnow Blewett heaped manuscript material at my feet. Floyd A. Paul opened the door to C. C. Lauritsen's past.

Friends and colleagues read and offered helpful suggestions on earlier drafts of this book: Don Anderson, Lee DuBridge, Tom Everhart, Jesse Greenstein, Willy Fowler, Bill Goddard, Norm Horowitz, Hiroo Kanamori, Ed Lewis, Hans Liepmann, Gerry Neugebauer, and Paul Spitzer. Ed Barber is a first-class editor and a very patient man.

My family deserves special thanks. Marcia and Mark Goodstein grew up patiently listening to tales of Caltech long ago, wondering when the saga would end. My husband, David, bore up, mostly cheerfully, kept me honest and to a schedule, and saw it through.

MILLIKAN'S
SCHOOL

ONE

◆

From the Orange Groves to "Camp Throop"

California is said to use the smallest matches in the world. It is also reported that the size of its liars is in inverse proportion to that of its matches. The theory is not so strange, since there are so many surprises and novelties in the conditions of California, especially the south end of it, when considered by the people of the Eastern States.

—NORMAN BRIDGE, 1899

ON NEW YEAR'S DAY, while the rest of the nation shivers in the cold, the annual Tournament of Roses parade gets under way in Pasadena, California. Dozens of flower-bedecked floats, colorful marching bands, and precision drill teams make their way down the city's principal thoroughfare, Colorado Boulevard. Thousands of people line the sidewalk, many of them having camped out in the same spot the night before. Millions more around the globe huddle in front of television sets, watching the Rose Bowl parade unfold before their eyes. On that day, the sun is invariably shining and the temperature is in the seventies, the palm trees sway in the breeze, and the snow-clad San Gabriel Mountains tower majestically in the near distance. More people decide on January 1 to move to southern California than on all of the other 364 days in the year put together.

The California Institute of Technology, perhaps the country's leading center for science and technology, is another Pasadena institution known round the world. It, too, has put on a show-stopping public spectacle from time to time. The most dazzling pageant in recent memory took place on October 30, 1969, on the occasion of the inauguration of Harold Brown as president of Caltech. For one thing, he was only the second president of Caltech, and the first one to be sworn in on the campus itself. The entire campus turned out to watch the investiture ceremonies.

Joined by several hundred delegates from other academic institutions and learned societies, Caltech's faculty fell into line and marched briskly down the grassy mall in front of Beckman Auditorium, their long academic robes and caps flapping in the breeze. The crowd leaned forward, hungry for a glimpse of Max Delbrück, professor of biology, and Murray Gell-Mann, professor of physics—Caltech's newest Nobel Prize winners. While Delbrück's award was already a week old, the Caroline Institute in Sweden had announced the news of Gell-Mann's prize the very morning of Brown's inauguration.

The Caltech band played, the glee club sang, and a host of invited dignitaries spoke to an overflow audience of guests and local citizens, who sat patiently for two hours under a scorching sun. It is doubtful that anyone remembered afterward what the speakers said on that particular morning. On the other hand, no one who watched the procession could forget how Delbrück and Gell-Mann, their faces beaming, danced and skipped and wiggled in line. The crowd beamed back at them. Such was Harold Brown's official welcome to Caltech.

There was a connection between that happy October morning in 1969 and the early stirrings of the American nation. It was the life of a restless American, who spanned a century and a continent, living and helping to shape that turbulent era between the end of the eighteenth century and the beginning of the twentieth. At the end of his life, the dreamer founded the institution that would go on to become Caltech: the link between the American past and its future, the East and the West. It is yet another version of a story that abounds in the annals of American history. In Caltech's case, the man's name was Amos Gager Throop.

Frontiersman, businessman, and philanthropist, Amos Throop lived and helped shape the American legend. Throop was born in upstate New York in 1811, but he spent most of his life exploring lands beyond the Hudson River. His heritage laid the groundwork of a fierce patriotism: the elder Throop, conscripted for the War of 1812, sent to serve his country the best substitute his money could buy. The restless family moved from place to place in New York State, but young Amos, like many young men of his time, sought a larger stage on which to play out his life. In 1832, at the age of twenty-one, he set out to seek his fortune in the wilderness of the Far West. He found it in St. Clair County, Michigan.

Before six years had passed—he spent them working in the sawmills along the Black River and rafting the lumber down to Detroit—Amos decided there were some touches of civilization he could not do without. He returned to New York to take a wife, a Miss Eliza V. Wait, with whom he returned to Michigan, and in 1843 they headed for Chicago.

It was in Chicago, where he lived for thirty-seven years, that Amos found the arena suited to his talents. A man of strong convictions, Throop was a politician to the very tips of his toes. He had been a Democrat until, still in Michigan, he found himself involved in a ballot box–stuffing swindle, whereupon he became an ardent Republican. He helped organize the Chicago Board of Trade, was elected alderman in 1849 and again in 1876, served as city assessor and city treasurer, spent five years as a member of the Cook County Board of Supervisors, and even spent a term in the Illinois state legislature, later calling it a "Sinque of corruption."

At one point, Throop barely missed being elected mayor of

Chicago on a Temperance ticket. Indeed, he ran for mayor twice on a Temperance ticket and once on the Republican ticket and lost each time. By conviction, Throop was an abolitionist. In 1850, he and some colleagues on the Chicago Common Council did battle with Senator Douglas over the Missouri Compromise issue. Douglas returned from Washington and made a fiery antiabolitionist speech at city hall, which prompted the crowd, unable to find Throop, to hang and burn his effigy in Dearborn Park. Amos once said that the lynching killed only the tree, but the episode was no joke. Soon afterward, Throop chaired a convention of the Old Abolition party in Springfield, Illinois. Abraham Lincoln refused to participate, finding their abolitionist views too strong for a man aspiring to public office. Fourteen years later, the staunch opponent of slavery lost a son in the Civil War.

Like Lincoln, Throop had been known to split rails during his youth in New York and Michigan. If it was not presidential timber, it was at least profitable lumber, and Throop multiplied the proceeds by means of real estate dealings in Chicago. A self-made man, Throop made and lost several fortunes in lumber and real estate. In religious matters, he was a Universalist preacher, one of the founders of the Second Universalist church, known as the Church of the Redeemer.

By 1880, when Throop once again headed west, starting a new life in Los Angeles, he was a wealthy man. He bought a farm in Los Angeles, near Jefferson and Main streets. He bought orchards and farmland, planted crops, built drainage systems, bartered and sold his produce, bought more land ($3,000 was the going price then for a ten-acre orchard), and generally prospered.

In 1886, Pasadena was a small but relatively sophisticated community of some five thousand midwestern transplants living among the orange groves. Throop first drove up by horse and buggy from his farm in Los Angeles to see if there was any interest there in holding Universalist meetings. Finding a positive response, he moved his family to Pasadena, founding the First Universalist parish of Pasadena (later renamed the Throop Memorial Universalist Church), with himself as moderator.

The aging Chicago politician and the ambitious young town suited each other. Elected to the city council in 1888, Throop became Pasadena's mayor one year later. Less than twenty years old, the town already had a public library and several public and private schools, but not a college. Other nearby towns had colleges: Pomona, Occidental, Whittier, and the University of Southern California were already established. Pasadena wanted a college. Amos Throop, then eighty years old, would provide one.

Throop had little formal schooling himself, and he seems initially to have planned a Universalist institution. Although it never had formal recognition from the denomination, it was to have nine Universalist parishioners from California on its board of trustees. In his diary for September 1, 1891, Throop noted how he had spent the day: "Planted potatoes, cleaned a water pipe, husked the corn . . . looked into corporation papers for school. In afternoon, saw Mr. Wooster and rented his block for five years . . . for our school purposes and hope I have made no mistake." Modestly named Throop University, the school opened its doors in Mr. Wooster's building in downtown Pasadena on November 2, 1891.

The university ran into difficulty immediately. In addition to the College of Letters and Science, there were to be courses in music, art, and elocution, preparatory courses, classes in stenography and typewriting, and a law school. But the initial faculty numbered only six, and everything was in short supply. The executive committee of the board of trustees appointed Throop himself chairman of a committee to find a janitor. He moved promptly, announcing at a meeting on November 7, 1891, that his nephew "has commenced this morning to take care of the building." More ominous was the report of the Committee on the Law School, again chaired by Throop. On January 12, 1892, the trustees were advised to "suspend all action in regard to the matter."

The great land boom of the eighties in Los Angeles had reached its peak in the summer of 1887. It had been a time of growing fortunes and limitless expansion for the region's newly minted academic community. The University of Southern California was preparing to build an astronomical observatory on Mount Wilson, just above Pasadena, when the boom collapsed. The observatory plan was abandoned so that USC could concentrate on the struggle to maintain its existence. An observatory on Mount Wilson was later to play a leading role in the fortunes of Amos Throop's institution, but 1892 was not the moment to create a grandiose university among the orange groves at the foot of the mountains.

Finding few university students in Pasadena, the school's trustees hastily revised the curriculum, now calling for courses that, in their words, fostered "a higher appreciation of the value and dignity of intelligent manual labor." (Schools specializing in such industrial training had come into vogue in late-nineteenth-century America, a reflection of the fact that most high school graduates went directly into jobs requiring manual or mechanical dexterity.) The only school of its kind in southern California to offer elementary school–through college-age students, boys and girls, training in the use of different tools and machines, Throop University barely survived its first year in Pasadena. In 1893, one year before Amos Throop's death, Throop's board of trustees voted to change the school's name to Throop Polytechnic Institute and built new quarters several blocks north of Colorado Boulevard.

In its first fifteen years, Throop Polytechnic became all things to all people, teaching a great variety of subjects, with considerable stress on manual training. In addition to launching a grammar school and a high school, Throop opened a teacher-training school, began a business school, and eventually added an elementary school to its collection of academic departments. In its quest for more students, Throop Polytechnic entered floats in the town's Tournament of Roses parade, an annual event since 1890, sent its Mandolin and Guitar Club on tour, and experimented with evening classes and summer sessions. By 1906, the six schools composing the Institute boasted an enrollment of 530 students, including 23 working toward the bachelor's degree in the College of Science and Technology.

At that point, Throop was working on its second president. The school had lost its first president, Charles Henry Keyes, over a matter of railroad bonds. The bonds, issued by the Mount Wilson Railroad Company, made up one-third of the school's endowment. By 1896, the railroad was on the brink of bankruptcy, and Keyes apparently wanted the school's trustees to dispose of its bonds before the railroad failed. But the Pasadena banker Paul Green, the president of Throop's board of trustees, and vice-president of the railroad, saw things differently. He persuaded the school board members to hold on to the bonds. Keyes resigned. Three years later, Throop cashed in its bonds, worth $19,500 at face value, for thirty cents on the dollar.

Walter Edwards, Keyes's successor, set about rebuilding Throop's financial base. In an effort to recruit more students, the school reduced its tuition by 25 percent. Its own students were also pressed into service. One Throop go-getter was dispatched north to Santa Paula, Ventura, and Santa Barbara to canvass for students. He received $7.50 toward the expenses of the trip, with a promise of $15 if he actually signed up a student. In another instance, Throop paid the tennis tournament expenses of a student who agreed to "talk Throop among the players and visitors."

The school also arranged for advertising in local newspapers and magazines. Not having money to pay for the ads, Throop typically offered one or more tuition scholarships in return for an equivalent amount of advertising space. It was good business for the newspapers, which ran free-scholarship contests as a way of attracting new readers. In the case of the Riverside Press scholarship, for example, the newspaper put up $50 cash, and $50 worth of advertising, in exchange for one year's tuition at Throop.

President Edwards also had Throop's accreditation to worry about. One examiner from the University of California in 1900 inspected the school and later reported that "the larger classes visited showed great lack of sobriety and discipline." The examiner didn't take too kindly to the English faculty either. "It is in charge of an old teacher who was long an incubus in the High School, but who was finally unloaded on Throop," he noted in his report. Suddenly, in 1905, Berkeley refused to accept course credits from Throop. When Edwards complained, Berkeley's president defended the action on the basis of past reports. When Edwards pointed out that Berkeley had taken the fees Throop paid for the service, but had not sent any examiners to Pasadena that year, Cal's president apologized, more or less, and promised a better review of Throop's credentials. The school passed the test the next year.

James A. B. Scherer was the last of the Throop presidents, arriving in 1908. Two years later, still beset with problems ranging from attracting decent college students to choosing between a coeducational and an all-male student body, Scherer decided to go ahead with construction of a new building on a new site, rather than lose public confidence in Throop. By then, the elementary school had left the Throop fold, moved to the southeast corner of Catalina and California, and become an independent private institution, now called Polytechnic School.

Throop's head owed his new job to the school's science-minded trustee, George Ellery Hale. New to the board, Hale had urged his colleagues to discard the high school and other programs and, in his words, "concentrate their entire attention . . . on mechanical and electrical engineering" and "adequate instruction . . . in the humanities." The trustees not only went along with the suggestion; they gave Hale the job of finding the right man to take over the school.

Hale would discover in the course of his search that none of the scientists he knew had the stomach for the job. One candidate declined, citing "the educational atmosphere of the Pacific Coast." A second had a long list of questions, including one about the composition of the board. "Under the old regime," Hale told him, "there is one woman on the Board, who is all right but will doubtless disappear in due process of time." She did. After he ran out of scientists to ask, Hale remembered a Lutheran minister he had met while traveling to Europe, in March 1907. On a hunch, Hale wrote to him and invited him to become Throop's president. Scherer, then head of a small sectarian college in South Carolina, said yes immediately.

Scherer had no formal training in science, few scientifically minded friends, and even less experience in organizing, staffing, and running an engineering school. But Scherer had a way with words, a flair for the dramatic, and a gambler's instinct. He raised the money to build Pasadena Hall (better known as Throop Hall), the first building on the present campus, pared the student body to thirty-one, hired Throop's first electrical engineer, and established the principle of an all-male school.

Scherer accomplished the latter by presidential fiat. "We have no women at all," he boasted to Hale and others. "I found that there would be so few that it would not be just to themselves to come here. Probably this settles the question of co-education, and I trust, without exciting the militant suffragettes." Hale, then abroad, took the news in stride. "I was much pleased to hear of the registration at Throop," he confided to Scherer, "and especially of those who didn't get in!"

The real test of Scherer's presidential mettle came from an unexpected source, the state of California. Politicians and educators wanted a technical school of higher learning in southern California, and they wanted the state to pay for it. In Sacramento, a bill was introduced early in 1911 to establish a "California institute of technology" in Los Angeles. That bill caused an uproar in Pasadena, where the city's new engineering college, Throop Polytechnic Institute, had only just moved from downtown Pasadena to a new campus located at California and Wilson, eliminated all schools except the college, and gone into debt over the purchase of extensive new equipment, mainly in electrical and mechanical engineering.

Throop's president Scherer had consistently opposed a University of California campus in southern California. He certainly did not need a rival technical school next door. The Sacramento bill called for an appropriation of one million dollars, nearly ten times Scherer's operating budget. If the bill passed, Scherer's private school had no future. He left for Sacramento immediately, determined to oppose the California Institute of Technology.

Scherer took with him, in fact, a counterproposal from Throop's trustees. The board, led by its president, Norman Bridge, offered to turn Throop over to the state of California. By week's end, a new bill calling for Throop Polytechnic Institute to become a state school had been introduced in the California legislature. "Throop stock has gone up 1000%," Scherer told a friend, and predicted the bill would become law within the month.

But Scherer, who had lived in the state for only three years, did not understand California politics. The University of California regents objected to the Institute's independent board of trustees, Cal's president objected to dividing the university's share of state funds with another institution, and the Berkeley alumni saw the Sacramento bill as a calculated plan by the southern alumni to gain a university.

Scherer was counting on the support of Stanford's president, David Starr Jordan. "It is certain," he had told Scherer, "that sooner or later southern California with nearly half the tax paying of the state will demand" and get a southern branch of the University of California. But under pressure from Berkeley, Jordan changed his mind.

On March 10, 1911, after endless hours of debate, the senate defeated the Throop bill. The Sacramento story was over.

But not Throop's. The press coverage alone, in California, bought Pasadena's pint-sized engineering school more public visibility than had all its previous efforts to canvass for students.

Scherer may have been Throop's president, but the trustee who had hired him, the solar astronomer George Ellery Hale, was the driving force behind the school's development in the years that followed. A leader in the science community, Hale had been elected to the National Academy of Sciences in 1902 and promptly set about to reform it. Back in Pasadena, Hale offered to help Scherer transform Throop into "a high-grade institute of technology," as he liked to put it. Hale proved an able tutor. They went abroad together, visiting astronomers and physicists, observatories and schools, in London, Paris, Rome, and other European cities. Closer to home, Hale introduced Scherer to other scientists, including a prominent physical chemist by the name of Arthur A. Noyes, of MIT, a close friend.

Like Hale, Noyes did not have to be asked twice to serve as an adviser to the new engineering college. He volunteered to help President Scherer in arranging courses, planning new buildings, and identifying promising young faculty in chemistry. In the spring of 1913, Noyes agreed to spend several months in Pasadena, looking over the situation firsthand.

The following year, Europe went to war. Fueled by Hale's determination to use the war abroad to advance scientific research at home, President Scherer threw himself and Throop into national defense work. In southern California, Scherer later boasted, no school outdid Throop College in "patriotic preparedness." Calling for "a new baptism of patriotism," the forty-six-year-old president spent six weeks in 1916 at a YMCA-sponsored military training camp in Monterey, California. "There are over 1000 here, including 10% of the Throop students and 25% of our Faculty," he reported to Hale. "I get up at 5 o'clock and keep at it steadily until too sleepy to sit-up longer; am hard and fit." At Scherer's insistence, school trustees voted to add military instruction to the curriculum, whereupon more than 100 of the college's 129 registered students petitioned school officials to make the training compulsory. "We granted their petition," the president rejoiced, already dreaming of turning the twenty-two-acre campus into one huge training camp.

A tireless crusader on behalf of preparedness, Scherer turned the war into his personal cause. A galaxy of speakers who shared his views came to speak at Throop, including such luminaries as

Arthur Twining Hadley, the president of Yale University; and James R. Garfield, who managed the Republican Charles Evan Hughes's unsuccessful 1916 campaign for the presidency. The students were treated to talks on topics ranging from "What a socialist thinks of pacifism" to "What this war means to us" and "What is a soldier?" As one student later recalled, "We were being keyed up to the necessity for preparedness. . . . [Scherer] was very, very much in favor of military preparedness as a necessity to preserving the Kingdom of God on the earth . . . [and] we all were sold on the idea of . . . defending everything that was good and holy in life." Even after America entered the war, the faculty committee on assemblies expressed concern that the students were "not well enough informed in regard to the fundamental conditions and ideals underlying the causes of the present war." Hale promised them a talk on this theme, later described as a " 'hummer' " by one listener, who added, "He has 'the goods,' if you will pardon so much slang, and no one can listen to him without being impressed." Unfortunately, Hale did not save the text of his remarks.

Intended to complement the college's engineering program, military instruction at Throop soon became an end in itself, threatening to eclipse the regular work of the college. Being an ordained Lutheran preacher, Scherer naturally linked Throop's cadre of student-soldiers to "the call of Christianity," as he phrased it. Passionate in his causes, he severed his ties with the Lutheran church in 1916 over its pro-German stand. But he did not abandon his spiritual tenets altogether. "The school is broadly Christian," he insisted, and continued to lead the students in saying the Lord's Prayer at every assembly. His religious and political convictions fused. "America has become," he confessed to an associate, "with all its implications, my religion." He viewed the military course in a similar vein, touting it as "a preparation for life, providing a sounder physique, giving alertness and quickness of action in emergency, and developing such high moral qualities as self-control and subordination to a common purpose in the unselfish service of society." Weapons were the one thing Scherer lacked.

Not for lack of trying, however. In August 1916, he applied to the War Department for full battle gear for 150 cadets. Turned down, he prepared a new, less ambitious shopping list: "I am wondering whether it might be possible for you to have the Ordnance Office send us immediately at least one complete uniform; one set of infantry equipment, complete with rifle; and fifteen of the new Springfield guns, latest model, which will be at the rate of one rifle for every ten students enlisted." Rebuffed a second time, Scherer tried a new tack: he asked the War Department to issue the college ten 1903 model rifles for use in target practice. Nothing came of this request either. The students made do without them. "We would get out there . . . in front of Throop Hall, out in the dirt, no lawn anywhere," a veteran of the college's drill brigade later told an interviewer, "[and] go through all sorts of exercises . . . using wooden guns." Months later, Scherer was still trying, without success, to obtain guns and other government supplies.

In January 1917, the army established a unit of the Reserve Officers' Training Corps, the Throop College Battalion, on the campus. In March, school officials began accepting applications for "Camp Throop," the training camp for 1,500 noncommissioned officers slated to open in mid-May on the grounds of the college. The trustees had approved this move, provided that the government furnished the necessary equipment and instructors, the college did not have to cover the added expenses, and the regular college program did not suffer. Tents and temporary buildings went up; a sewage system was installed—local businessmen put up the money. School closed a month early in anticipation, and preparations mounted despite growing rumors, discounted by the president, that no federal appropriation existed for camps such as Throop's. "In spite of the Adjutant General and the Secretary of War, Camp Throop will be held," Scherer boasted to a friend, a week before the three-month course in infantry, cavalry, engineering, and sanitary instruction was scheduled to begin.

This fine talk and white tents dotted the campus. But Scherer's euphoria was short-lived. The next day, he learned "that the Adjutant General would not even permit the Western Department to issue Krags for our use, although they had quantities of these old unused guns in their possession." The last straw had landed. Scherer ordered the tents hauled down. "We were not willing to have an unreal or 'fake' camp," he explained.

Such an abrupt collapse of the camp unsettled Scherer. He flirted briefly with a campaign for the governorship of California on the GOP ticket and inquired about a State Department job in the Far East. In the end, he followed Hale and Noyes to Washington in June 1917 and settled for a dollar-a-year job working as chief field agent for the Council for National Defense, leaving the vice-treasurer, Ned Barrett, in charge of Throop. Government proved so congenial to Scherer that he took an indefinite leave of absence from Throop. An outspoken public critic of William R. Hearst's newspapers, Scherer resigned his government position one year later, in protest over Secretary of War Newton D. Baker's banning speakers from making "discriminatory remarks as to the relative values of newspapers." Never one to do things halfway, he also published an open letter to the war secretary in the newspapers, condemning the Hearst papers for selling a "German peace" to the American public. Hale congratulated him and told him to come home.

Scherer returned to campus to find that the college had contracted to train three hundred men under the government's newly organized Students' Army Training Corps (SATC) program. "The students are rolling in on us, and it keeps us moving to 'keep ahead of the game,'" he observed. "Hammers din in my ears as I write, while through my windows I can see the roofed mess hall and raftered barracks, while the huge tent that we have secured for temporary accommodations looms up white and big on the graded area once prepared for tennis courts." In all, three barracks and a mess hall were constructed at government expense. Two hundred and forty students were inducted into the SATC unit. The War Department contracted with the college for the students' tuition and for room and board. Uniforms and other equipment were also provided. Scherer had finally got his weapons.

On the morning of October 1, 1918, nearly three hundred inductees and students gathered in front of the campus flagpole. Scherer welcomed them with a stirring, patriotic speech. "Soldiers of the United States Army," he began,

> students of Throop College of Technology: This is the brightest day that the world has seen for more than four years. Not since the first of August, 1914, has the sun of civilization shone as it is shining today. Bulgaria out of it, Turkey staggering, Austria-Hungary tottering, and Germany cold to the marrow with fright. . . . America has won a new self respect. Old Glory shines with a new luster. Our boys "over there" . . . are fighting as though the war were to be over by Christmas. We here must prepare as though it were to last twenty years.

Forty-two days later, however, the war was over. The SATC was soon legislated out of existence, and by December 1918, soldiers were being demobilized directly from the college. It seemed that Scherer's dreams were dashed, but not entirely, for peacetime ROTC units sprang up on college campuses across the nation, including Throop's.

The educator Scherer's Christian crusade for America, and his war games on campus, threatened Throop's development. Tailoring its own curriculum to the government's program, Throop had already gone the full route. It had abolished modern-language instruction, added a course on the causes of the war, and eliminated others in history and economics. On top of that, the SATC men had been sent to work in the kitchen and in the commandant's office, filing papers. Collegiate work suffered, in part because the heavy military load cut into the normal academic schedule, in part because many of the students were primarily interested in military training. "We have had no *college* spirit," Scherer groused, ". . . and there has been a military spirit, which is all right in war time but is not so fine nor uplifting." But he refused, even so, to abandon military training at the college. The one bright note about the SATC program came from the director of physical education, who observed "that military training had put . . . [the football team] in splendid physical condition."

In less splendid condition was the academic life at Throop. Rumors began to circulate that many students, particularly the freshmen, were planning to withdraw from the school at the end of the first semester in 1918. Like others on the faculty, Edward ("Ned") Barrett, business agent, vice-treasurer, and de facto president of Throop during Scherer's wartime absence, feared for the college's future. Choosing his words carefully, Barrett urged Scherer to put the war behind him and behind the school.

> . . . I believe there may be a danger of losing men who really did want a course like ours, and who could stay and take it if they wish to. This danger may come from the desire on the part of the men to be free from the military regime reaching over into a desire to get away from everything that has been connected with it, including the college classes. . . . It is, of course, clear that the men have got very little out of their college work this term.

Talk to the freshmen, he pleaded with Scherer; "tell them what Throop is, what our courses are, and especially of your determination that the military work in the future shall not take so much time as to prevent genuine college work." Within days, Scherer had talked personally with the undergraduates. Anticipating a 25 percent drop in enrollment, the college notified high schools of its intention to admit "properly prepared students" to a midyear class, starting in February.

Despite Barrett's misgivings, enrollment had climbed steadily during World War I. In 1914, attendance at the college jumped from 58 to 91 students; in 1915, it reached 129, rising to 185 the following year. When the student population dropped from 155 to 135 during the first semester of 1917–18, the college admitted a new freshman class in February 1918 (a practice continued until 1921), raising the total for the year to 193. Operating under a year-round schedule during the war, the college graduated two classes in 1918, leaving no seniors on campus in 1918–19. "We have 230 students actually in attendance," Scherer wrote one trustee in 1919, "and the outlook for next year is almost depressingly promising. I seriously contemplate recommending to the Board [of trustees] limiting the attendance to 300 hand-picked men."

Regardless of the number, Scherer's talk of limiting the college's enrollment raised fundamental questions about its future. Extrapolating from the growth in enrollment since 1910, Barrett estimated in 1916 that the number of students admitted to the college would reach 500 by 1921. He made up a budget for the college; it did not include the cost of new buildings or the cost of doing research. Figuring on a faculty of seventy, an average professorial salary of $2,045, and modest outlays for upkeep and equipment, he projected $200,000 in annual expenses. In 1916, the cost of attending the college was $150; even at that rate, tuition for 500 students would bring in only $75,000, far less than the projected costs of educating a student five years from then. Where would the remaining $125,000 come from?

From endowment. The arithmetic was simple enough: 5 percent interest on $2.5 million. The college then had, according to Barrett's records, "an endowment fund for general support of $500,000." It needed $2.0 million additional endowment, if his calculations were correct. Who would provide it? In late 1916, Arthur Fleming became president of the college's board of trustees. Like Scherer and other school officials, Barrett assumed that Fleming, a self-made millionaire, would provide it.

By the turn of the decade, Throop's enrollment stood at 337 students. Scherer thought the time had come to put a cap on the student body, because the assembly room had reached its capacity. Noyes thought the time had come "to remove the name Throop and . . . attach to the Institution the name of the great state of California." Hale thought the time had come for Throop to become the scientific school of the future. "It is merely a question of policy and men," he wrote.

In early 1920, at one of his regular weekly student assemblies,

Scherer broke the news that the college had changed its name to the California Institute of Technology. Someone from the back of the room called out, "Hooray for Caltech!" "Oh please, don't abbreviate it," Scherer pleaded with his audience. His words were destined to be ignored.

TWO

•

Preamble to a Technical School

You ought to devote yourself to the inauguration of
great plans.
—A. NOYES to G. Hale,
October 15, 1907

GEORGE ELLERY HALE came to Pasadena in 1903 to establish a solar observatory on the summit of nearby Mount Wilson.
Founded in the 1870s by a group of settlers from Indiana, Pasadena then boasted an extensive interurban electric railroad system, coast-to-coast service on the Southern Pacific Railway, a
public bikeway and library, 125 automobiles, and round-the-clock
telephone service for its 15,000 inhabitants. In the course of borrowing money, leasing land, buying lumber, renting pack animals,
and hiring architects, carpenters, and stonemasons, Hale set his
stamp on the young city.

In the thirty-five-year-old Hale, the townspeople found a scientist bursting with educational, architectural, and civic ideas.
Fond of telling others to "make no small plans," Hale himself
sparked the creation of the Huntington Library and Art Gallery,

worked on the master plan for Pasadena's civic center, joined the board of trustees of Throop Polytechnic Institute, and played a major role in transforming it into the California Institute of Technology.

It all began with a box of tools. As a young man growing up in Chicago, Hale haunted the local astronomical observatory. His father, a wealthy elevator manufacturer, instilled in young George a love for tools and machinery and a deep interest in public affairs. Hale's mother, a graduate of the Hartford Female Seminary, in Hartford, Connecticut, cultivated his literary side, reading aloud the *Iliad* and the *Odyssey* and stocking the shelves of his personal library with books ranging from the unabridged *Robinson Crusoe* and *Don Quixote* in translation to *Grimm's Fairy Tales* and the poetry of Shelley and Keats. In biographical notes written years later, Hale spoke fondly of these and other classics that "helped greatly to arouse [his] imagination and prepare [him] for scientific research."

Armed with a box of tools and a small lathe for turning metal, Hale transformed his bedroom into a laboratory and later constructed with his own hands a workshop in the yard. The aspiring scientist built first a small telescope and then a spectroscope. Equipped with these instruments, Hale observed the pattern of light and dark lines in the solar spectrum. Reading everything he could find on spectra during these years, Hale bought Norman Lockyer's *Studies in Spectrum Analysis*. Intrigued, he began laboratory observations of spectra and compared them with those of the sun's spectrum. The exercise turned the amateur astronomer into an astrophysicist. In explaining why classical astronomy, with its emphasis on determining the positions, distances, and motions of celestial bodies, did not appeal to him even then, Hale later wrote, "I was born an experimentalist, and I was bound to find the way of combining physics and chemistry with astronomy."

Determined to leave his mark on the young science of astrophysics, Hale entered the Massachusetts Institute of Technology (MIT) in 1886, where he studied chemistry, physics, and mathematics. During summer vacations, he continued his own solar and stellar research in a special spectroscopic laboratory of his own design, built for him by his father on a lot adjacent to the family home in the Kenwood section of Chicago. This formed the nucleus of the Kenwood Physical Observatory, his first observatory.

At the time, the standard method of recording solar prominences—gaseous eruptions on the sun—was to make drawings on the basis of visual observation. In his quest to find an adequate method of photographing prominences, Hale invented the spectroheliograph—an instrument for photographing phenomena in the solar atmosphere that would otherwise be invisible. The first tests of his new instrument, made in the winter of 1889–90 at the Harvard Observatory, demonstrated that the basic principle was right. Then a senior at MIT, he wrote up the work for his thesis and received a B.S. in physics. Hale married Evelina Conklin in June 1890, two days after graduation. Upon the couple's return to Chicago, Hale's father agreed to finance the construction of a 12-inch refractor telescope.

At the Kenwood Observatory, dedicated in 1891, Hale continued his experiments with the spectroheliograph. In examining the spectra of prominences, he observed two particularly bright frequencies (H and K), which he had determined to be due to calcium, in the ultraviolet region. They proved ideal for the photographing of prominences, as photographic plates then in use were more sensitive to light in the ultraviolet. Moreover, his photographs of solar spectra showed H and K to be bright all over the sun. Armed with this information and an improved instrument, Hale photographed these calcium clouds (flocculi) and prominences at both the solar limb and across the disk in 1892 for the first time. The success of this research tool launched Hale's career.

In 1892, he joined the faculty of the new University of Chicago as associate professor of astrophysics, bringing with him the Kenwood Observatory. An accomplished organizer and money-raiser,

Hale then persuaded the streetcar millionaire C. T. Yerkes to provide Chicago with the largest refractor telescope in the world, to be housed in the university's new Yerkes Observatory, which Hale would direct. Before he was thirty, Hale had designed and built one world-class observatory and founded an international scientific magazine, the *Astrophysical Journal*.

The catalyst for Hale's move west was a newspaper announcement in 1902 that Andrew Carnegie had earmarked $10 million to establish an institution devoted to pure research. Hale spent the next two years lobbying the Carnegie Institution of Washington trustees for funds to build a solar observatory at Mount Wilson. Savoring the scientific advantages of the peak above Pasadena—a good site and ideal climatic conditions, meaning astronomers could count on "good seeing" (the astronomical term for peak visibility) over an extended period of time—honed Hale's talents for raising money. He himself later explained, "I was thus bound to undertake the heavy task of raising funds or to forgo the possibilities I seemed to see ahead. These were nothing less than an effective union of astronomy and physics, directed primarily toward the solution of the problem of stellar evolution, but with equal consideration of the advantages to be gained by fundamental physics from such a joint study." Confident that the Carnegie Institution would underwrite the Mount Wilson project, Hale settled his family in the fall of 1903 in Pasadena. The Carnegie trustees turned down his request that December.

Undaunted, Hale went ahead with his observatory plans, staking $30,000 of his own money on the project. In June 1904, the owners of the Pasadena and Mount Wilson Toll Road Company, hoping that Hale's observatory scheme would boost tourist traffic to the summit, signed a lease granting him a portion of the peak, rent-free, for ninety-nine years. Technically, their new tenant was simply on an expedition as director of the University of Chicago's Yerkes Observatory. But Hale doubted that the university would establish another observatory. A gambler at heart, he put his own name on the document. He added a clause reserving the right to transfer the lease to the Carnegie Institution, should it decide to establish a solar observatory of its own in southern California.

His gamble paid off in December 1904 when the Carnegie Institution authorized $300,000 for the establishment of the Mount Wilson Observatory and appointed Hale its director. A solar astronomer primarily, Hale also built stellar telescopes: a 60-inch reflector installed at Mount Wilson in 1908 (which Andrew Carnegie came in person to visit two years later) and a 100-inch telescope inaugurated in 1917.

Faced with the need in 1905 for a permanent instrument shop and optical laboratory to service the observatory, Hale appealed to Pasadena's board of trade, receiving a pledge of a suitable site in return. Before long, the board's shopkeepers, doctors, and lawyers had raised $1,400 in subscriptions toward the purchase of a lot costing $2,100 on Santa Barbara Street. The new brick laboratory, completed and equipped by the end of the year, became the office of the Mount Wilson Observatory.

Affluent, socially astute, and politically conservative, Pasadena's leading scientist mingled easily with the town's other prominent citizens. Membership in the Twilight Club, a convivial private club to which Throop's president at the time, Walter Edwards, belonged, and the sports-minded Valley Hunt Club, the exclusive preserve of Pasadena's class-conscious, high-society families, gave Hale a head start in his new surroundings.

Pasadena earned the reputation early in its history of catering to people with a lot of old money. There was a popular saying at the time that went, "Rich people who move to Southern California do not go to Pasadena to live unless they have had money for at least two decades." In his autobiography, the Throop trustee Norman Bridge, a resident of the city between 1894 and 1910, tried to soften its popular image as an upper-class oasis of plenty. "During those years," he recalled,

Pasadena got the reputation of being a small city of millionaires. Really the rich people were few; the large numerical majority were poor people, living in little inexpensive bungalows, most of which were wrongly constructed for protection of their occupants against the heat of a few hot days of summer, and the nights following such days.

Hale's own contacts with Pasadena's less privileged citizens were limited. Japanese laborers were employed in the building of a road to the mountaintop observatory, and he watched their progress daily from his home, with the aid of an 8.5-inch telescope. He had acquired the house, a big, rambling, vine-covered structure in South Pasadena, surrounded by five acres of open fields, spreading trees, and lush gardens, from his personal physician, James McBride. McBride, who happened to be a trustee of Throop, had swapped his home for Hale's in Pasadena. The money for Hale's way of life came mainly from dividends and interest on stocks and bonds, which in 1910 exceeded $12,000, four times the salary of the highest-paid professor at Throop. While Hale was not a millionaire himself, his prestige gave him access to those who were.

When he found himself seated next to Henry E. Huntington, the electric railway car magnate, at a local banquet in 1906, Hale seized the moment. He told Huntington, a collector of rare books and paintings who lived in neighboring San Marino, about his solar observatory and his plans to turn Pasadena into a research center and, in Hale's own words, "pointed out how admirably the creation of an exceptional library would harmonize with this scheme." It took another twenty-one years, a common interest in the Pasadena Music and Art Association, which Hale helped found, the presence of the California Institute of Technology, and Hale's spirited objections to Huntington's original plan to turn his San Marino estate into a public museum, but in the end Hale's scheme for the Huntington Library and Art Gallery—a research institution, incorporated and managed by a private board of trustees—came to pass, largely as he had envisioned.

By 1907, he had begun to take an interest in Throop. A man

who disliked public speaking and who generally shunned the limelight, Hale nevertheless agreed that year to make a public address in Pasadena on Throop's possibilities. Speaking in the school auditorium that January, Hale appealed to the townspeople to make right what was wrong with the education of engineers elsewhere. Engineers in college typically lived on a steady diet of "specialized technical courses," he explained, remembering all too well the undergraduate experiences of the engineering students he had known at MIT. A better way to train engineers had to be found. Throop's "prime object should be to graduate men capable of conceiving vast projects," an impossible task, he added, without "a broad scheme of education which may give proper recognition to all sides of the engineer's life." A member of the local board of trade—Hale seemed to belong, indeed to direct, nearly everything—Hale carefully aimed his remarks at the pocketbooks of the doctors, lawyers, and businessmen in the audience, pointing out that southern California couldn't survive, let alone grow, without engineers to bring water and electrical power to the semiarid region.

Who would train these engineers? Hale ruled out the schools on the East Coast because of their distance, and those on the West Coast—the University of Washington and Stanford University, among others—because of their uneven quality. In any event, the nearest engineering school was located in northern California, in Berkeley. To Hale, the simplest solution lay close at hand. "Under such conditions, and with the advantages afforded by climate, by the immediate neighborhood of mountains where water-power can be developed and experimental transmission lines installed, who can deny that there is a place in Pasadena for a technical school of the highest class?" he asked the group. Intended to define the policy of the school and to draw attention to the potential of an engineering school in southern California, Hale's public address in 1907 had a more fundamental impact on the pursuit of science in America between the two world wars. In Pasadena, his dramatic question turned the spotlight on the smallest of the six schools composing Throop Polytechnic Institute, the College of Science and Technology. What emerged was not entirely encouraging. Although Throop's collegiate department boasted courses leading to a bachelor of science degree in electrical engineering, chemistry, or biology, only twenty students had earned Throop diplomas between 1896 and 1907, causing one school official to declare, "We could and should do more." In the beginning, the college had no prescribed courses, only electives. Specified course work leading to a bachelor of arts degree in chemistry, electrical engineering, physics, and natural science came later, in 1900. Three years later, physics fell by the wayside. By 1905, the college had blossomed into the "College of Science and Technology." A school bulletin the next year referred to it as the "College of Technology and Science," suggesting perhaps that school officials considered them to be of equal importance.

Scholarly credentials were also in short supply. No one teaching science had earned a doctorate; indeed, no member of the faculty possessed a Ph.D. In chemistry, Wallace Gaylord, a graduate of MIT backed by fourteen years of classroom experience and postbaccalaureate work at Berkeley, taught both inorganic chemistry and qualitative and quantitative analysis. The only member of the physics department, Professor Lucien Howard Gilmore, educated at Stanford and the University of Chicago, taught physics and electrical engineering. A thirteen-year veteran in the classroom, Gilmore also served as the school's inspector of equipment. In addition, Throop offered instruction in mathematics, biology, and mechanical engineering, the last subject the domain of Robert Ford, an alumnus of the University of Minnesota, who also ran the school's manual-training program.

Therefore, Hale's 1907 lecture drew a swift response from school trustees. Uncertain for some time of what course to pursue, the trustees turned to Hale that spring for advice. Elaborating on his public talk, Hale suggested they abandon trying to teach everything "in a mediocre way" and concentrate on doing "some one thing extremely well." When they asked him that spring what kind of school he favored, Hale replied, a "high-grade institute of technology."

If it was left up to him, Hale said, he would not duplicate other technical schools but rather "educate men broadly and, at the same time, make them into good engineers." His remarks mirrored those contained in a provocative article, "A plea for the imaginative element in technical education," which he wrote in 1907 for an MIT magazine. In this article, Hale offered his own prescription for change at his alma mater. Decrying the temptation simply to fill their students' heads with equations, he urged school officials to provide more, rather than less, instruction in the humanities. In the name of the unity of science, Hale proposed a course of lectures on evolution, ranging from the origin of stars and the Solar System to the origin of continents and mountains and the origin of species and society itself. "If science is to be regarded as not inferior to the humanities in its educational possibilities," he declared, "it is because it deals with the largest and most fruitful conceptions, of which evolution is perhaps the greatest." Instead of more time for specialized technical work, he preached the need for more general studies. As an antidote to compartmentalized course work, he called for thematic science museums. Finally, he even called for adapting the architectural design of the buildings to the purposes of the school. While intended as a blueprint for change at MIT, Hale's paper found a more positive and immediate response in Pasadena.

Two announcements later that year sealed Throop's future. In June, the Institute issued a bulletin describing the requirements for admission and an outline of the courses to be offered in its "College of Technology and Science." Seeking, in their words, "to build up a technical school of college grade the equal of any in the country," the trustees announced their intention to expand and develop the college along engineering lines, starting with courses in electrical engineering. Demand for well-trained engineers on the Pacific Coast in general and in the Southwest in particular merited the school's new emphasis. Required and elective course work in nontechnical subjects would provide "an education broad in its scope and deep in its intensity." In August, Hale formally joined the Throop trustees.

Within a short time, the Throop elementary school broke away and set itself up as an independent, private institution called Polytechnic School. One by one, the remaining schools—the grammar and high school, the normal school, the commercial school—closed their doors, casualties of Throop's gradual conversion into a college. The trustees had agreed to keep the high school going for another two years, on a trial basis. But by that time there were twenty other vocational high schools within commuting distance of Pasadena. In 1911, Pasadena voters approved a $475,000 bond issue to establish a public polytechnic high school in Pasadena. That spring, Throop's trustees voted to end the operations of their high school.

At Hale's invitation, James A. B. Scherer began his presidential tenure at Throop in 1908. Under its new leader, Throop turned its emphasis to electrical engineering, something no other college or university in southern California could boast. Plans called for constructing an electrical engineering building first, to be followed by separate buildings for the other branches of engineering. As scheduled, the first building on the new campus of Throop Polytechnic Institute served as a laboratory for Royal Sorensen, who gave up a job in the engineering department of the General Electric Company in Pittsfield, Massachusetts, to start a department of electrical engineering at Throop in 1910. Long on plans and short on funds, the entire college—civil, mechanical, and electrical engineering, chemistry, physics, languages and English, administration, library, and bookstore—was housed in Throop Hall (known then as Pasadena Hall) for some years.

As college president between 1908 and 1920, Scherer tackled the school's deficit, hired and fired instructors, wrote books and gave speeches, and increasingly sought a larger political arena for himself. Untutored in science, he dealt with the day-to-day problems of running an educational organization, including raising money for the school. The broader problems associated with putting into practice the school's new approach to technical education fell to Scherer's mentor, Hale—"working out the educational scheme," Hale called it. In the beginning, the division of labor suited both parties. "Some of the things I could do very well, but others would be much better done by another man," Hale once said, in explaining why he had avoided becoming a college president himself. If he shunned that office elsewhere, he ended up, nevertheless, in the driver's seat at Throop on a number of occasions. One of the most important was his campaign to secure for the fledgling university the services of the celebrated chemist Arthur Amos Noyes.

In the life of every great institution, there is a moment when fortune smiles so broadly that it lights the future. Such a moment occurred in the spring of 1913 when George Ellery Hale journeyed back east for a stay with one of his old professors from MIT, Arthur Noyes. Would Noyes be willing to spend several months the following year at Throop College? Hale had posed the same question before, but this time Noyes's answer was different. He found the invitation to winter in Pasadena appealing and, after talking the plan over with MIT's president, Richard C. Maclaurin, told Hale that he would like to rent a small bungalow during his stay in Pasadena. Although Noyes's commitment was quite limited—a mere two months—once back in Pasadena, Hale began to talk as if Noyes would make the part-time arrangement a permanent one. Noyes, taking a wait-and-see attitude, suggested trying it out for one year first. Of course, the temptation to come and to stay would be much stronger if Hale could produce, say, a chemistry building. "I don't know where this [building] is coming from," Hale admitted to his wife, cautioning her not to say anything just yet about the visit. To Hale's surprise, President Scherer talked Charles W. Gates, a member of the board of trustees, into providing seed money for the new building. Gates, a retired businessman living in South Pasadena, pledged $25,000

toward construction of Gates Laboratory, the first science building on campus. On the strength of this promise, Noyes, the dean
of America's physical chemists, "Arturo" to his closest friends,
boarded the Twentieth Century Limited in Boston in late January
1914, bound for Pasadena. Forty-seven and unmarried, Noyes was
tugged by loyalty to MIT after twenty-five years on the job, but
he was frustrated with the school's approach to educating engineers. He had decided to come to see for himself what Hale and
Scherer were up to in Pasadena.

Behind Noyes's visit lay Hale's determination to rebuild
Throop from the ground up into "a technical school of the first
class." The educational philosophy of the two scientists, who had
become close friends during Hale's MIT student days, meshed
perfectly. In Noyes, Hale saw a chance not only to bring chemistry
at Throop up to a level with that at MIT but also to put Throop
itself in the national limelight. By the time war had begun in
Europe in 1914, nothing less counted for Hale. By weaning Noyes
from MIT, he just might succeed.

Noyes was the charter member of Throop's scientific colony.
He had spent years preaching the need to reform the education
of America's engineers. Stressing the place of physics, chemistry,
and mathematics in the engineering curriculum at MIT, he had
placed a premium on teaching students "principles rather than
specific industrial applications." Without the former, he told a
group of MIT freshmen in 1907, "you would be only rule-of-
thumb engineers, who could imitate, but not initiate." That same
year, he became MIT's acting president and immediately set about
trying to put his ideals into practice. But his insistence that fundamental scientific study was essential to creative work in engineering challenged not only the prevailing wisdom among
practicing engineers but the thinking of colleagues on his own
campus as well. Whereas he pushed for more undergraduate
course work in the basic sciences, the engineering rank and file

promoted the study of basic industrial techniques as a way to train chemical engineers. Whereas Noyes called for more laboratory work coupled with practice in solving, as he put it, "problems . . . that develop logical thinking or reasoning power and the kind that develop imaginative thinking or the power of planning and originating," the traditionalists, like the MIT chemist William H. Walker, stressed on-the-job problems.

The clash between Noyes and his MIT colleagues over the ideal system of education in applied science spilled over into other academic areas. Like Hale, Noyes advocated educating engineers broadly. Art, literature, music, and history figured prominently in Noyes's definition of a technical education. Much to his dismay, the various engineering faculties at MIT drew the line at course work outside of their own branch of study. Unable to reform his engineering colleagues, he stepped down after two years as acting president, tired and depressed by his ineffectual crusade.

Meanwhile, an offer from the University of California at Berkeley sat on his desk. In 1908, UC Berkeley's president, Benjamin Wheeler, intent on upgrading the state university's graduate work in science, had offered him the chairmanship of the chemistry department. Wheeler looked to him, as Noyes later recounted, to "give tone to the department." For Noyes, the new job promised a respite from his obsession with MIT's problems and future. "I almost think," he confided to Hale, "it would be for the interest both of my scientific work and of my peace of mind for me to go to some other institution, at any rate for a time. . . ." Confident that Hale could get the information easily, Noyes pressed him for an opinion about Berkeley's academic standing. "Is it," he asked with more than a trace of East Coast snobbism, "still so undeveloped physically and intellectually, as to make a position there relatively unattractive, and to make it impracticable to develop a strong graduate and research department in chemistry?" Don't even consider Berkeley, Hale told Noyes.

According to colleagues there whom he trusted, the "research spirit" was "by no means dominant, or even widespread among the faculty" at Berkeley, Hale informed Noyes. He admonished Noyes to remain in Boston, "in the midst of the scientific life of the country." No sooner had he advised Noyes to stay put than Hale began courting Noyes himself. Despite Hale's reservations, Berkeley's shortcomings as a research institution in the early part of the century paled by comparison with Throop's.

"Unknown and still undeveloped" was Hale's own candid assessment of Throop Institute. It did not stop him from formally inviting Noyes in 1909 to become the head of the chemistry department at the Institute as well as dean of the engineering school. "You are the only man in the country, so far as I am aware," Hale declared in 1909,

> who understands fully the difficulties and possibilities of such an undertaking. . . . If you chose, you would be given a free hand to develop the Engineering School; on the other hand, if you did not wish to undertake this work, you would be at liberty to devote yourself entirely to chemical research, simply giving us the privilege of discussing with you the problems encountered in working out the educational scheme. What we want is your presence on the ground, in any capacity you might choose to occupy.

In essence, Noyes could write his own ticket at Throop.

The invitation struck a sympathetic chord in Noyes. Citing his duty to stand by MIT, he declined Hale's offer (and Wheeler's too) but left the door open for a future visit to Pasadena. Short of joining the faculty, he was willing to do what he could to build up Throop.

By then, Throop College of Science and Technology had discontinued the practice of granting bachelor of science degrees in chemistry and biology. Reflecting its new emphasis on applied science, the school's course work in 1909 for electrical and mechanical engineering students, substantially the same for both groups, culminated in a B.S. in engineering. Along with these changes, the college's enrollment had risen to thirty-four. Keen to corral Noyes, Scherer begged his advice in hiring faculty. Throop needed a descriptive and analytical chemist: Did Noyes know someone? Could Noyes recommend an instructor—"some alert and rising young man" in physics and chemistry? What about an experienced industrial chemist? Preferably an outgoing person, Scherer said, since much of his time would be spent interesting local oilmen in the work of the school. "Personality always counts with us for at least fifty per cent," he added, in elaborating his personal criteria for staffing the school, "and in this is, of course, included good breeding." While sympathetic, Noyes could think of no one who fit that description.

In the meantime, Hale promoted the Throop connection whenever he went east. Between meetings and visits to Noyes's summer retreat in Maine, Hale kept the Pasadena invitation alive. In the summer of 1912, Noyes spent part of his summer vacation out west, visiting Hale in Pasadena. Scherer showed him around the campus. Hale saw Noyes again in Boston the following March and, once again, invited him to visit Throop.

This time, Arthur Noyes was definitely interested. As Hale reported to his wife in mid-April 1913, Noyes would come in exchange for a new laboratory. Other matters relating to the visit also had to be settled. Scherer played no direct role in these negotiations. In luring Scherer to Pasadena, Hale had all but guaranteed him success as a money raiser. "This is a very wealthy community, and the conditions for raising large sums of money appear to be unusually good," Hale had told him in 1908. But even after five years of hard work, Scherer had little to show for his efforts. In the course of working together in Pasadena, Scherer and Hale became good friends; so did their families. Under the circumstances, Hale was anxious that he alone be the one to break to Scherer the news about the laboratory. After four years of singing Scherer's praise to Noyes, and Noyes's to Scherer, Hale preferred to keep the news about his old friend under wraps until he knew the full price of the arrangement.

The details of the arrangement, worked out as the two scientists traveled together to attend the National Academy of Sciences meetings, reached Scherer's desk in Pasadena two weeks later. Hale had promised Noyes not only his own chemistry building but also a course in chemical engineering, a salary of $1,200 for the period February 3 to March 28, 1914, and the services of a temporary assistant. In return, Noyes agreed to teach one term of inorganic chemistry and an introductory course in physical chemistry, to train others to carry on the teaching in his absence, and to overhaul Throop's chemistry curriculum. At the end of April 1913, on the eve of his departure from New York for Europe, Hale finally disclosed these details in a letter to Scherer. Eager to put the best light on the string of promises he had made in Scherer's name, Hale reminded Throop's president, "Noyes would also give us the great benefit of his advice on the work of research. This alone would make his coming very desirable."

Scherer trusted Hale's judgment completely. (And he was apparently a selfless man.) If he felt eclipsed when Hale took so much into his own hands, he kept it to himself. Scherer, in any event, was already on good terms with Noyes, had toured his laboratory, and spent time in the Noyes home. Besides, Noyes did not covet Scherer's job.

Noyes had his own preoccupations. A scientist who knew his way around research laboratories on two continents, Noyes had spent more than a decade drilling the research spirit into promising MIT undergraduates and staffing his own laboratory with research-minded chemists. An avid sailor, he had even given his thirty-foot yacht the name *Research*. MIT had in fact been Noyes's life. After taking an undergraduate degree there, Noyes stayed on to earn a master's degree in chemistry as well; his reward was a junior teaching appointment for the 1887–88 academic year. He later wrote of his first MIT teaching experience, "Had full charge of class of 30 or 40 men in qualitative analysis. Was a strenuous year as I had no adequate knowledge of the subject. Had as students George E. Hale [two years younger than Noyes] and Harry M. Goodwin [later a physicist at MIT] and became very intimate with them." The year up, he went abroad for advanced study. By the time he received his Ph.D. degree, in 1890, from the University of Leipzig, in Germany, Noyes had worked in Johannes Wislicenus's organic chemistry laboratory for a year and discovered that he preferred physical chemistry and Wilhelm Ostwald's physical chemistry laboratory, where he spent the second year carrying out research on the solubility of salts.

On his return to MIT in 1890, Noyes taught, did research, rose through the academic ranks, wrote textbooks, and trained graduate students—indeed, the first crop of MIT's own Ph.D.'s carried out their doctoral research in his laboratory. He founded the laboratory in 1903 and personally contributed half the cost of its maintenance every year. Independent of the chemistry department, the Research Laboratory of Physical Chemistry, which consisted of twelve rooms spread over two floors, housed a library, offices, stockrooms, and research laboratories, including one set aside for undergraduate use. The building stood within earshot of a heavily used railroad line. At Throop, Noyes clearly expected more and better facilities.

Amid these impending changes in Pasadena, Throop officially changed its name to Throop College of Technology in April 1913.

President Scherer secured Charles Gates's pledge in 1914. But even in pre–World War I America, $25,000 did not buy a chemistry building, much less equip it. With no other pledges in sight, the laboratory was put on hold; Noyes himself canceled a return visit in 1915. At first, Arthur Fleming, vice-president of the board of trustees and the school's chief financial backer, attributed the problem of financing the building simply to a downturn in the economy. Then came the war in Europe in 1914. Convinced from the start that the war would last a long time, Fleming advised Scherer to curb the school's expenses. "While the country," he informed Scherer from his business headquarters in Boston, "or at least the middle and eastern portion of it, is full of money, people seem unwilling to separate themselves from it because of the uncertainty of the future." Efforts by Scherer to secure the funds for the proposed building from the steel magnate Andrew Carnegie failed, as did a similar proposal by Hale to the Carnegie Institution of Washington. Having exhausted these and other leads, Scherer and Hale reopened talks in Pasadena in 1915 with Charles Gates.

Talking sparked an idea. Remembering that his brother Peter had included a bequest for Throop in his will, Charles Gates sounded him out about putting up the cash in return for a lifetime annuity. Peter was less than enthusiastic about Charles's plan. Knowing of his brother's interest in antiquity and Hale's in Egyptology, Charles next suggested a meeting between the two men. Hale agreed to the visit, promising, in Scherer's words, to "avoid the danger of getting P.G. so much interested in archeology that he will lose track of the importance of the chemistry building at Throop." Still, May came and went without any announcement about the building.

Undeterred, Hale brought a trustee into the picture: Arthur Fleming. Elected to the board in 1903, several years after he and his wife had moved to Pasadena, the Canadian-born lumber tycoon had taken the school under his wing, lavishing both money and time on it. Until their deaths in 1904, his wife, Clara Fleming, who assumed in 1898 a mortgage on the school's property, and Clara's father, Eldridge Merick Fowler, also included Throop among their local philanthropic causes. By 1907, Fleming had become indispensable to Hale's vision of the school's future.

Fleming was simultaneously a generous benefactor of Throop Institute and a relentlessly frugal trustee. Whereas he paid Scherer's annual salary out of his own pocket for years, Fleming was notorious for his small economies. He chided Scherer for raising faculty salaries in 1915, claiming—incorrectly, as Scherer pointed out—"the Professors at Throop have had much larger pay than the Professors of any other College in the state." Fleming did not hesitate to provide the twenty-two-acre site for the new school campus at Wilson and California, at a cost of $50,000; he also personally covered its annual deficit on more than one occasion, and contributed handsomely to its first building, Pasadena Hall. At the same time, and not without reason, Fleming often acted as if he owned the place. Deeply enmeshed in the business and financial management of the Institute, Fleming took a personal interest in all aspects of his investment, from shrubs to payroll. At one point, he objected to landscaping the campus with Italian cypresses, on account of their "cemetery appearance." But when Hale finally went to him in 1915 to secure the money for Noyes's research laboratory in physical chemistry, Fleming fell into line without a murmur.

Insisting on anonymity, he promised to give Throop Institute $10,000 for equipment, and an equal amount annually for salaries, including Noyes's, and other expenses. He might be the school's principal benefactor, but he did not want the public to think it had no others. The offer, conditional upon Noyes's agreeing to settle permanently in Pasadena, put the chemistry building and Noyes's annual visits to Throop back on track. It was Hale's show from the start. While on the East Coast for a round of meetings, Hale had stopped off to see Noyes and Fleming in Boston. On May 5, 1915, the three men had breakfast together at Noyes's house. "Can't tell how it will come out," Hale wired back to his wife and President Scherer, breaking the news to them about Fleming's offer to Noyes. Calling it "Throop's superlative opportunity," Scherer sent Noyes a telegram from Pasadena endorsing the plan and urging him to accept Fleming's offer. Three days later, Hale set out for home, pausing long enough in Chicago to send two night letters of his own to Noyes, pressing him to make a clean break with MIT. Eager to know all the details of the "Boston episode," Scherer met Hale's train on the outskirts of Los Angeles and rode with him the rest of the way. "The acceptance of Noyes is practically assured," he told Scherer. It seemed too good to be true. "Personally, I feel very small and humbler than usual (usually I am not very humble)," Scherer exclaimed. "I feel deeply complimented that a man like Dr. Noyes should be willing to come here with me as President. If his coming depended upon my resignation I should gladly offer it, so much importance do I attach to his acceptance." Noyes, however, no longer had any presidential aspirations: his experiences at MIT had taught him otherwise. The challenge to Scherer's office came instead from an unexpected quarter: the war in Europe, which altered the balance of power at Throop. For the time being, though, Scherer was content and his job secure.

To Hale's chagrin, Noyes wanted only to loosen his ties with MIT, not cut them—at least not yet. Convinced that his research laboratory at MIT would be swallowed up by the chemistry department if he left, Noyes offered to divide his time equally between Throop and MIT for two years, as an experiment. Hale protested. "The plan I presented to Fleming," he wrote Noyes,

> was not primarily to establish a research laboratory at Throop, but to bring you, with all of your influence . . . to assist in building up the school and the intellectual life of the whole state. . . . Think how scientific and educational men would have been impressed by the actual . . . possibilities of Throop if we could have announced that you had wholly cast in your lot with us. . . . It would put Throop "on the map" at once and help enormously in taking the next similar step.

But Noyes stood his ground. He was coming primarily to develop research work at Throop, and until his new building had been constructed and equipped, very little research was going to get done. Fleming himself had not only agreed to the half-time plan but also suggested to Noyes that it might be in Throop's best interest to have him allied with both schools.

Then there was the matter of how much time Noyes would give Throop. Hale and Scherer wanted the equivalent of half of the school year, four and a half months, so that Throop could say Noyes was dividing his time equally between the two institutions. Noyes, who had allotted three and a half months to Pasadena and almost four to Boston, replied that his responsibilities to the graduate students in his laboratory at MIT outweighed his responsibilities at Throop. "I can see no reason whatever why you should not announce that I was dividing my time practically equally between the two institutions," he wrote them in May 1915. When Noyes also said that he could not put the half-time plan into operation immediately, Hale advised him by return mail to stick with routine research in Boston, and work out new lines of research in Pasadena. "I do consider it *very important* that you should come next winter [1916], and that some research should be started at once under the new auspices, mainly because of its effect on Fleming." Noyes proposed, instead, a brief visit later in the year. Hale did not press the point ("the main thing we wanted was to be able to announce that you would give half your time to Throop") and pronounced himself satisfied with the arrangements.

At commencement exercises that June, Scherer announced that Noyes would give half his time to Throop. By the end of that summer, the funds for the chemical building were in hand. Charles and Peter Gates each promised to provide one-half of the cost. The New York architect Bertram Goodhue had been hired as a consulting architect on the building and the general plan of the campus. Noyes had lined up two young chemists, Stuart Bates and Howard Lucas, to take charge of the details of equipping the new building and had begun to plan the research work. Attendance at Throop in 1915 had climbed to ninety-one.

In December 1915, during his promised visit to the school, Noyes addressed the students:

> There are in America at present two important centers of scientific investigation—the endowed research laboratories, like those of the Carnegie Institution or the Rockefeller Institute on the one hand, and the universities on the other. Of these the universities are by far the most important; and if Throop is to be a scientific and engineering school of the highest type, it cannot be content merely with educating, however effectively, engineers and chemists, but it must do its part in contributing to the progress of American science.

His speech must have pleased Hale, for in it Noyes left little doubt that he intended to play a major role in shaping the school's

future. In declaring that Throop "must do its part in contributing
to the progress of American science," Noyes breathed new life
into Hale's vision of a technical school. In insisting that Throop's
duty also lay in contributing to the advancement of scientific
knowledge, he gave a new twist to Hale's view of Throop as a
training ground for engineers. In summing up his own educational ideals, he told the students,

> . . . industrial research is not the main research opportunity of ed-
> ucational institutions, and it is not the side on which their important
> contributions to science have been found in the past. Industrial
> research will in the end take care of itself. . . . The main field for
> educational institutions is research in pure science itself—a study of
> fundamental principles and phenomena, without immediate refer-
> ence to practical application. . . . Scientific investigation is the spring
> that feeds the stream of technical progress, and if the spring dries
> up the stream is sure to disappear.

Given his battles with MIT officials, it is perhaps not surprising that Noyes spoke out as strongly as he did on the place
of science in higher education. Driven by his belief in a broad
scientific school, where science and engineering are treated like
equals, rather than a narrow technical school, where the fundamental sciences are treated like stepchildren, Noyes resolved to
succeed at Throop where he had failed at MIT. That was the real
significance of his assembly talk in December 1915.

If Noyes had done nothing else in Pasadena that month, his
visit would have served its purpose.

Noyes's visit set other wheels in motion. Pleased with the
establishment of the chemical research laboratory, Noyes now
called for school officials to do the same for physical research.
Like chemistry, physics would require a building and a research
fund and a research-oriented faculty. Scherer's school lacked all
three. "I most enthusiastically share your view that Throop should
be a great school of science, perhaps even a university in time,"
Hale had told Noyes earlier in the year, "and you may count on
me to do everything in my power to bring this about." Hale

couldn't resist tweaking him a bit: "Of course this is the very reason I wanted to get you here."

In characteristic fashion, Hale and Noyes started with the faculty. They had little difficulty in identifying the scientist needed to place physics on the same plane as chemistry at Throop. In discussing "who this man might be," Scherer reported to Fleming that December, "We all agree that the most eminent man in the country is Millikan of the University of Chicago. . . . He is like Noyes in several respects, being a capable and enthusiastic instructor as well as a great research man." Would Robert Millikan come? "Our bringing Noyes here of course makes it easy to get other distinguished men to come," bragged Scherer, who was counting on Fleming to find the necessary financial backers for such a plan. From the start, neither Hale nor Noyes had any doubts about their ability to get Millikan to come to Throop on a part-time basis. On this optimistic note, Noyes returned to Boston in time to welcome in the new year, 1916.

MIT never seemed the same again to Noyes. "I have become much more *warlike* since the fear of being unduly influenced by your persuasive influence has been removed," he wrote Hale upon reaching home. He began to talk more openly of leaving MIT; he wrote more belligerent letters to its president about the Institute's imbalance between science and engineering course work, but he couldn't bear to leave the place. It was another four years before Noyes could finally bring himself to resign from MIT and accept a full-time appointment in Pasadena.

It had taken three years to raise the $70,000 needed, but on March 10, 1916, ground was broken on the Throop campus for the Gates Laboratory of Chemistry. What had begun in 1913 as a conversation between two friends in Boston now grew quickly into a two-story, reinforced-concrete building containing thirty-nine rooms, research laboratories, and a basement, the exterior tan-colored walls set off by carved stone and wrought-iron trim around the windows and the main entrance. Nine months later, on December 1, 1916, Noyes officially took up his part-time duties as director of chemical research at Throop College.

THREE

♦

The Birth of Caltech

My dear Scherer—

Back from Washington after an interview with the President and some other activities. I am made chairman of a National Academy committee to organize the scientific resources of educational and research institutions in the interests of national preparedness. In the event of a break with Germany the Carnegie Institution will make all its facilities—men and equipment—available for service as far as needed. Probably most of the regular work would be continued, but men especially qualified for work bearing on military affairs would give their time to it. I will ask you later to secure such an offer *through the National Academy* (as in the case of the C.I.) from Throop.

—GEORGE E. HALE,
April 26, 1916

EUROPE'S WAR came as a great shock to most Americans, most of whom wanted nothing to do with it. As the newspaper editor of one midwestern daily told his readers, "We never appreciated so keenly as now the foresight exercised by our fore-fathers in emigrating from Europe." From the White House to the War Department and the average citizen in the street, the spectacle of eight countries killing, maiming, and laying waste to historic cities was reason enough to remain aloof. American tradition, President Woodrow Wilson declared in 1914, demanded neutrality, not American intervention overseas. Swept into office the year before on a mandate to reform the country's business and financial affairs, the Democrat Wilson threw himself into domestic issues, leaving others to worry about foreign entanglements. One reporter later wrote that "the first phase of Wilson's attitude toward the war was distaste for it, wish to keep it at arm's length, as something shameful to the world, odious to him personally."

But in this area, as in so many others, George Ellery Hale was the exception. Convinced from the start that the United States would eventually abandon its position of neutrality, the prescient Hale had begun as early as spring 1916 to organize and recruit scientists from university and industrial laboratories across the land to work on military problems. After several months of intense lobbying within and outside the government, Hale's plans for an organization to support such work, operating under the auspices of the National Academy of Sciences, culminated in the creation of the National Research Council.

By 1916, however, Wilson had also accepted the idea of military preparedness, and was pushing Congress to increase the size of the standing army. Nevertheless, the president was not yet ready to ask for universal military training for all young men, as some preparedness leaders wanted. In a speech in Pittsburgh that January, Wilson told the audience, "This country should prepare herself, not for war . . . but for adequate national defense." Congress resisted, but in the end passed the National Defense Act in 1916, which increased the troops to 175,000 men. Throughout 1916, Wilson persisted in thinking he could help negotiate an armistice between the Allies and the Central Powers, without American military intervention. The news given to President Wilson in January 1917 that Germany intended to resume unrestricted submarine warfare on February 1 brought the United States to the brink of war—and Hale's efforts into full flower.

As founder and first chairman of the Research Council, Hale lived in Washington for much of the war. While championing the immediate need to make the United States scientifically and technologically independent of countries overseas, he also wanted to advance research in physics and chemistry in educational institutions at home. Backed by President Wilson's pledge to him in spring 1916 "that the entire field of research belongs to us"— that is, to the National Academy—Hale set about organizing the council, raising private funds for its operation and outmaneuvering his critics. "I must be present to defend the radical departure from Academy traditions which the Research Council represents," he told Scherer, apologizing for his prolonged absence from Pasadena.

Geared to the needs of the military during the war, the Research Council tackled problems ranging from the manufacture of nitrogen compounds for the production of explosives to the building and testing of submarine detection devices and the physiology of battlefield shock. By June 1916, Arthur Noyes, a charter member of the council's organizing committee, was in Washington briefing the secretary of war, in his official capacity as chairman of the nitrate supply committee. In case of war, Hale confided to an associate, he wanted to put the University of Chicago physicist Robert A. Millikan in charge of all research work. "I really believe this is the greatest chance we have ever had to advance research in America," Hale boasted in a letter to Throop's head, forging a link between his own efforts in Washington and President Scherer's in Pasadena to promote research in the interest of preparedness.

To Hale, World War I was the best thing that could have happened to Throop. He lobbied not only for science to play a larger role in national affairs but also for Throop to play a larger role in American science. In fact, Hale used the war shamelessly to promote the transformation of Throop College of Technology into the California Institute of Technology.

The resolution Hale offered on April 19, 1916, at the close of the National Academy's annual meeting in Washington had been on his mind for months. It came on the heels of the torpedoing in March of the *Sussex*, a French channel steamer, off the coast of France. In the event of a diplomatic break with any nation, Hale called on the academy to offer its services to the president. On the same day that President Wilson publicly condemned Germany's actions, Hale's resolution carried by an overwhelming vote. Before the meeting ended, the cosmopolitan astronomer had also been reelected foreign secretary.

It was also Robert Millikan's first annual academy meeting, and he happened to be sitting directly behind Hale when he rose to speak. Greatly in favor of Hale's resolution, Millikan rushed up afterward and "told him so with vigor." Shortly thereafter, Hale returned the compliment by appointing Millikan a member of the Research Council's organizing committee.

Although Millikan wrote a number of letters from Washington to his wife, Greta, in Chicago, he hardly mentioned this meeting of the National Academy. Thrifty to a fault, he did grumble about having to spend five dollars to attend the banquet. At the time, Millikan was president of the American Physical Society. To hear him tell it in 1916, the only memorable event at the NAS meeting was the paper he heard on the sex of a parthenogenetic frog. The physiologist Jacques Loeb's results, it would appear, had taught the forty-eight-year-old father of three sons a lesson in biology. Referring to Loeb's technique of artificial parthenogenesis, in which a needle prick triggers embryonic development in an unfertilized egg, Robert wrote to Greta, "He finds that he can only get *males* that way." Continuing in this vein, he added,

"You see any old pin-prick is enough to start off a man but a woman is still a *mystery. I knew it before!!* and I will tell you some more things which I have known all along as soon as I get home." Traveling alone, as Millikan liked to remind his mate, was no fun.

One thing is certain: Hale thought big. A week later, on April 26, 1916, Hale and four other members of the NAS, including President William H. Welch and the Carnegie Institution of Washington head, Robert S. Woodward, met with President Wilson at the White House. After Welch spoke of the academy's historic role as scientific adviser to the government, Hale presented his proposal of a war role for the academy, stressing the importance of scientific research in wartime and the need to include science in any plan of preparedness. "We must not prepare poisonous gases or debase science through similar misuse," he told Wilson, "but we should give our soldiers and sailors every legitimate aid and every means of protection." After asking a few questions, Wilson told them "to go ahead with the work, war or no war." Fearful that a public announcement would only worsen relations with Germany, Wilson cautioned them to keep their activities confidential.

Armed only with the president's verbal request, Hale spent the next five weeks putting together committees and lining up the cooperation of government bureaus, professional societies, and schools, as well as of prominent scientists and engineers. Shortly after his meeting with Wilson, Hale outlined his strategy to Scherer:

> Now my plan is to get every institution where research is done to feel that it ought to contribute toward some work of investigation for national service—men, money, laboratories, apparatus, etc. In these days of preparedness many a board of trustees, if hitherto indifferent, should begin to realize for the first time what research may mean to the nation. . . . With Noyes at the head of chemistry— in a movement which should go on indefinitely—you can see the interest of Throop in the work.

Money, as always, lay at the center of the scheme.

In principle, Hale remarked once, he supported the idea of federal appropriations for science; in practice, he took exception to a bill, introduced (but never passed) in the Senate in 1916 by Senator Francis G. Newlands, of Nevada, which would have provided funds for engineering stations at land-grant colleges. Since he lacked faith, on the one hand, in "the average American" to understand the importance of science and was convinced, on the other, that public-supported university research would inevitably end in "hardening of the arteries," Hale had also taken it upon himself in 1916 to raise a $1 million national research fund from private sources. "The spirit of national service in the air, coupled with the desire for preparedness, should make everything possible," Hale told Scherer that May, adding, "And I am going to give Throop the chance to lead the way." Initially he wanted the trustee Arthur Fleming to offer $25,000 through the school to the National Academy. The funds would be available for research, not necessarily at Throop. The academy would act as broker, and the school would benefit from the publicity. Unsure of Fleming's reaction to such a plan, Hale assured Scherer that the money could also be used at Throop, "though a little more latitude," he admitted, "would sometimes be very desirable." Along the way, it turned into a $100,000 request.

Eager to demonstrate that philanthropy was the answer to government support of research, Hale pressed Scherer to sound out Fleming. Practical considerations played a role as well: "As long as there is no fund to attach it [the National Research Council] to the Nat. Acad. there is danger that men like Cattell, representing rival societies which are down on the Acad., may demand an independent organization." James Cattell, a prominent psychologist, in particular, had accused Hale of using the war to buy the NAS a monopoly on science; Hale had replied that it was the only national organization with the potential, in his opinion, to perform "a national service of the highest kind." High or low, Hale was counting on Fleming's help to silence his critics, to launch the academy's national preparedness campaign, and to set an example for other schools. If Fleming insisted on giving the money directly to Throop, then at least Throop's meager endowment would increase.

Matters came to a head on June 6, 1916, at a gala banquet celebrating the twenty-fifth anniversary of the University of Chicago. Hale was scheduled to speak at 8:00 P.M., central time, on the topic "Scientific Research for National Service" and planned to announce the formation of the National Academy's research fund. The guest list included John D. Rockefeller and other members of the Rockefeller Foundation, AT&T's chief engineer, J. J. Carty, and other wealthy and influential men. A week before the speech and with no donations in hand, Hale had appealed to Scherer's civic pride: "If I could say on . . . [that] occasion that Throop had led the way in starting a National Service Research Fund it would be a splendid thing. . . . Southern California has no great reputation as yet for preparedness, but this would put it on the map."

Hearing nothing, he sent Scherer a telegram from Chicago on June 3; and another, the following day. Back in Pasadena, Scherer was waiting for Fleming, who had been out of town. He returned on Monday morning, June 5, in time for school commencement. On June 6, shortly after 7:00 P.M., Chicago time, Scherer telegraphed Hale. Fleming's positive response, Hale later insisted, came as no surprise, even if it did arrive barely thirty minutes before the banquet began. The college had also pledged to place its research facilities at the disposal of the National Research Council. "This is great business," Scherer later told him. "I want you to know that the Trustees were influenced not only by the opportunity to serve the country and science and the college, but also by the desire to 'back' you."

Less than two weeks later, the council of the NAS approved the organization of the National Research Council. Hale had promised Scherer, who was beginning to lose his enthusiasm for raising money, suitable recognition. "Throop has been given full credit in my speeches, and will be shown as the leader as soon as we can publish," he told Scherer, who was still waiting for a public announcement at the end of June. On July 29, 1916, under pressure from Hale, the White House issued a press release. That same day, the first public story about the establishment of the Research Council also appeared in the *New York Times*, under Hale's byline. Of the school's role in launching it, he wrote,

> Throop College of Technology, in Pasadena, California has recently afforded a striking illustration of one way in which the Research Council can secure co-operation and advance scientific investigation. This institution, with its able investigators and excellent research laboratories, could be of great service in any broad scheme of co-operation. President Scherer, hearing of the formation of the Council, immediately offered to take part in its work, and with this object, he secured within three days an additional research endowment of one hundred thousand dollars.

The new funds were earmarked for physics.

Although Gates Laboratory was still under construction in July 1916 and no graduate students were on the Throop campus (the chemist Roscoe Dickinson was to earn Caltech's very first Ph.D., in 1920), none of this deterred the enthusiastic Hale from promoting the school's modest research facilities. At the same time, Fleming's gift of $100,000, which was real enough, set in motion a chain of events that would lead to the establishment of a renowned physics facility, the Norman Bridge Laboratory. Hale had kept his part of the bargain with Scherer and, perhaps more important, had secured the down payment needed to lure away from the University of Chicago none other than Robert A. Millikan.

Shortly before Hale had received Millikan's enthusiastic endorsement at the April National Academy of Sciences meeting, he and Arthur Noyes had paid a visit to the physicist's home base in Chicago. Unaware that his name had already been floated in Pasadena as a worthy addition to Throop's scientific colony, Millikan set aside a day that April for Hale and Arthur Noyes's visit; they toured his laboratory, talked about his oil-drop and photoelectric work, and looked at what his students were doing. This was no ordinary courtesy call. Millikan had never met Noyes, although he had heard the name mentioned often enough in connection with the chemistry program Noyes was starting up in Pasadena. Hale he knew only slightly better; letters Millikan addressed to him during this period still began, "Dear Mr. Hale." But by the end of their next meeting, at the end of the April 19 NAS gathering, Millikan had agreed to attend the first meeting of the National Research Council scheduled for September and to take charge of organizing the work of the NRC in physics. Save for discussing with Millikan their plans for him in southern California, Hale and Noyes were batting a thousand. "I haven't broached the Throop scheme to him yet, but am only waiting for a favorable opportunity," Hale wrote to Scherer. Recalling his visit to Chicago's Ryerson Physical Laboratory, the ever ambitious Hale added, "If we should even get him for half the year Throop would have some part in the direction of American research!"

The first favorable opportunity came on July 18, 1916, in New York City, where Millikan was teaching summer school at Columbia University. After a day of conferences, Hale took Millikan to dinner at the Biltmore Hotel. The next morning, Millikan could not wait to tell his wife what they had talked about the previous evening: "He is the most *restless* flea on the American continent—more things are *eating him* than I could tell you about in an hour. Among many other things he wants me to spend 3 months of each year in Pasadena starting a new physical laboratory there." They talked again in August; Millikan refused to commit himself.

With an offer from Columbia also in his pocket, Millikan could afford to take his time. Columbia's offer included a salary of $7,000, which was $2,000 more than he was making at Chicago, and a $1 million endowment fund for research in physics. After talking it over with Noyes, Hale countered with an offer of $2,500 for one quarter annually in Pasadena—double the amount Millikan earned in three months at Chicago. If Millikan turned Columbia down, Hale wrote Scherer, Throop stood a good chance of getting him. In the meantime, there was much work to be done: Millikan would need a physics building, suitable research laboratories, equipment—certain to be "much more expensive than the chemical equipment"—and a larger research fund than Fleming had just provided. By October 1916, Millikan was definitely interested in the Throop offer. "I think I can arrange to join you in your Pasadena enterprise," he wrote Hale. What academic quarter would be best for him to come west? Knowing that Noyes was planning to spend the following winter in Pasadena, Hale proposed the same schedule to Millikan. Columbia's physics faculty refused to consider the matter settled; it offered Millikan more money; he turned a deaf ear. "I dread telling them that it's no use," he told his wife, Greta.

Appointed director of physical research at Throop College, Millikan gave the first of six public lectures on physics at the college on January 31, 1917. But America's entry into World War I put an end to any immediate talk of the University of Chicago physicist's dividing his time between the two institutions. On February 3, the United States broke diplomatic relations with Germany; forty-eight hours later, Hale appointed Millikan chairman of a special committee to deal with submarine detection problems. With three lectures still to go, Millikan took the overnight train up to San Francisco that same day, hastily took notes on underwater work in the Bay Area, and returned to southern California on the seventh, in time to speak on the subject "Brownian Movements and Sub-Electrons." Millikan lingered in Pasadena for two more days, just long enough to finish the course of lectures: "The Structure of the Atom," followed by "The Nature of Radiation."

Washington beckoned. "If the science men of the country are going to be of any use to . . . [the country], it is now or never," Millikan declared upon his arrival there at the end of March 1917, plunging directly into his new duties as vice-chairman, director of research, and executive officer of the National Research Council. On April 2, President Woodrow Wilson delivered his war message to Congress, and Millikan, a Republican to the core, almost wished he had voted for the Democratic incumbent. War or no war, perfecting Hale's organization, he calculated, would keep him busy on the East Coast for several months, no more. Millikan quickly found out otherwise. Between being a member of a special submarine board of the navy, chairman of the optical glass committee of the War Industries Board, and the head of the science and research division of the Army Signal Corps, Millikan did not see the inside of his Chicago laboratory again until the war ended in November 1918.

Heavily involved in the affairs of the Research Council and the National Academy at that point, Millikan lingered in the East for almost another year. By then, Noyes had already returned to Boston and Hale to Pasadena. Resuming his academic duties at Chicago in October 1919, Millikan also launched an ambitious new research program in hot-spark spectroscopy. This field he had begun to develop, he reminded Hale in 1919,

> just before you flew off to Washington with me dangling from your claws two years ago. . . . Perhaps you recall that I was working on a preliminary notice of the method in our office in Washington and when you overheard me dictating it, you turned and asked what I was doing with what sounded like a scientific article.

Millikan's wartime administrative duties did not go unrewarded. Indeed, few among the corps of scientists engaged in research for the military during the war years gained as much influence on American science in the postwar decades as Millikan did.

The Hale-Millikan-Noyes alliance, born in prewar Pasadena and nurtured in Washington during the war, was cemented in 1921 when Millikan left the Midwest for a new, full-time career in southern California. Working together during World War I, Hale later wrote, "paved the way for our subsequent cooperation in developing the California Institute." The school's initial Washington-MIT connection, cultivated by Hale, Millikan, and Noyes, brought the scientific talent. Throop marched off to the war as a school of engineering and applied science. It returned as the California Institute of Technology, taking its present name in 1920.

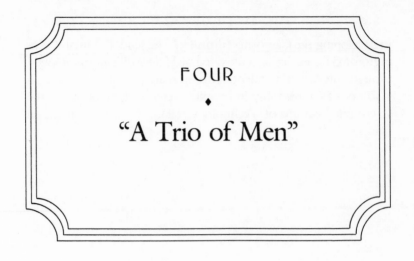

FOUR

◆

"A Trio of Men"

The great work of the NRC was initiated and developed by a trio of men who had vision, purpose, and inflexible determination to be of service to their country and to humanity.

—CHARLES D. WALCOTT,
1924

AT THE CLOSE of World War I, the structure of scientific research in America waited to be determined. Research in Europe and Scandinavia took place mainly in free-standing institutes. France could point to its Institut Pasteur and the Radium Institute; Germany had a chain of Kaiser Wilhelm Institutes; Denmark was about to establish the Institute of Theoretical Physics. In the United States, the National Research Council (NRC) had won its spurs organizing research for the war effort. Not quite a government agency, it was not entirely private either. The great private foundations, especially the Rockefeller and the Carnegie, stood ready to promote scientific research if someone would show the way. The traditional seats of American research—private universities such as Johns Hopkins, Chicago, and Columbia, and public universities in virtually every state—stood ready to answer the call. Throop College, in Pasadena, the dusty site of recent war games, would not have seemed a natural choice to join such an elite save for a central fact—the three men who had initiated and developed the NRC were determined to put Throop on the map.

Collectively ambitious for American science, eager to see their country play a larger role on the world's scientific stage, and ever mindful of their little institution in Pasadena, Hale, Millikan, and Noyes had become a formidable scientific triumvirate by 1918. Not that these three were short on critics. "There are many who believe," one observer noted, "that the Research Council is merely a device through which the National Academy hopes eventually to corner science in this country." Detractors faulted Millikan for his "exuberance," Noyes for his "fanaticism," Hale for minding everyone's business but his own.

Meanwhile, back in Pasadena, the main order of business seemed to be speculation about the net worth of the lumber magnate Arthur Fleming. How much money did he have? How much was earmarked for Throop? Fleming volunteered few details. But by Armistice Day his interest in the college verged on obsession. He "buttonholes every man he meets with the expectation of arousing his enthusiasm for an institution he has never seen," George Hale reported to the school's president, James Scherer, in Pasadena, following a round of meetings with Fleming in 1918 in New York, adding, "He seems to suppose that his own tremendous interest in the college must be shared by everyone, and he cannot understand why they do not at once turn over to him sums ranging from one to five millions."

How best to support America's scientific enterprise was a question on many minds as World War I drew to a close. In February 1918, George E. Vincent, president of the Rockefeller Foundation, wrote a letter to Professor Robert Millikan, in care of the NRC's Washington office. He wanted Millikan's opinion about the merits of endowing a single, independent "research institution to deal with physics and chemistry." Would such an institution, he wanted to know, provide the "leadership in American progress in the physical sciences" in the postwar era? Could a new or existing agency of the federal government do the same job better? Were there plans to turn the NRC into a permanent organization? Vincent's question about a research laboratory organized apart from a university got Millikan's full attention.

"I have hardly known how to reply and wanted to stimulate my own thinks through *argumentation*," the dutiful husband wrote to Mrs. Millikan afterward, describing a lively debate at breakfast several days later with Willis Whitney of General Electric and others over the merits of Vincent's research laboratory. "When I am particularly argumentative," Millikan reminded her, "you will know that it is only for the sake of forming my own opinion, not for the sake of putting an opinion over."

While her husband of sixteen years served in Washington, Greta Millikan, back in Chicago, was spending her time on the domestic front teaching knitting for the war effort to other faculty wives ("I have become a curiosity since I've learned to knit two socks at the same time"), fussing over her eldest son, Clark, and his two younger brothers, Max and Glenn, and bemoaning Robert's extended absence from home. She chided her husband in one letter, "I can see that with Mr. Hale hors de combat and the Council grown to the proportions it has, you cannot consider such a trivial thing as a visit to your girl and kiddies."

Indeed, the NRC was very much on Millikan's mind. In his reply to Vincent's letter that same February, he brushed aside the notion that the Bureau of Standards or any other federal organization could take the lead in building up American science. A dyed-in-the-wool conservative, Millikan put little stock in the idea of expanding the role of government in science, insisting that "an institution which is free from all political influences is better adapted to stimulation of research in pure science." Although the NRC operated on the fringes of the federal machinery, a point that Millikan did not dwell on, its future looked good; most important, "any scheme of stimulating research could be made to fit into its program." At the same time, the seasoned college professor took a dim view of pouring money into research laboratories not connected with universities.

What is the best way to do science? Millikan next asked a number of other scientists. Calling it "still an unsettled question" in his mind, Millikan nevertheless saw advantages to doing science in an academic setting. "The founding of a new institute," he went on to tell Vincent,

> like the Kaiser Wilhelm Institute at Dahlem [in Berlin] is the most direct and obvious solution, but I am not sure it is the best one.
> . . . The enlargement of vision which comes from the combination of research, with a certain amount of instruction, and from the broadening of one's contact with men, is unquestionably an aid to research itself. I have some half formed ideas with regard to the possible modification of the plan, which I shall be glad to submit to you in the near future. . . .

But before his own "half formed ideas" could succeed, Millikan had to win over Simon Flexner, who figured prominently in the workings of the Rockefeller Foundation, and George Ellery Hale, who had recruited Millikan for his own NRC post in Washington.

Teeming with ideas for his organization, Hale spent the winter of 1918 organizing committees, drafting legislation, and writing proposals aimed at preserving as well as enlarging the NRC's sphere of influence in national and international affairs. By the end of March, Hale knew that some members of Vincent's foundation, including Vincent himself, favored turning over a chunk of money to the NRC for a fixed amount of time, leaving it to the council to choose how to use the money to build up physics and chemistry. Flexner, however, wanted Rockefeller to run its own institute. Indeed, Vincent's initial correspondence with Millikan had grown out of these discussions. As Flexner's personal choice to head the new Rockefeller institute, Millikan was wined and dined at the physiologist's Madison Avenue mansion in New York. As a scribbled note to Greta following one such visit suggests, Millikan took his social obligations seriously: "I haven't done any of the four things I told you last night I might do . . . instead, it being raw and wet, I . . . went up to Simon Flexner's and, he being away, had a perfectly delightful chat with his deaf wife for a full hour, I regret to say, but she was too interesting to make it possible to get away earlier." Charmer though he was, Millikan never did persuade Helen Thomas Flexner's husband to abandon the central, independent research laboratory idea.

Hale, however, was beginning to see things Millikan's way. Of his ongoing discussions with Hale, helped along by the fact that they shared an office in Washington, Millikan wrote to his wife that they were, as he modestly called it, "planning the future of American science!!" The upshot of their planning was a proposal to the Rockefeller Foundation, which the two of them discussed with Vincent in New York that March. Hale described the plan in a letter to the Throop trustee Arthur Fleming:

> I shall accordingly propose to the Rockefeller Foundation that it provide the Research Council with funds for the establishment of research laboratories in connection with three educational institutions—one in California, one in Chicago, and one in the eastern part of the United States.

Although Hale stopped short of identifying the other schools—Arthur Fleming could read the University of Chicago and MIT between the lines as well as anyone—he used the occasion to put the touch on Fleming: "Colonel Millikan [Millikan's war efforts had earned him the rank of colonel] and I feel that the best place for the California headquarters would be Throop College of Technology, provided that its actual endowment and its laboratory equipment were great enough to warrant us in recommending it in preference to Berkeley or Stanford." He continued,

> You have several times spoken, both to me and Dr. Scherer, about your plans for the future of Throop. Would it not be feasible for

you to turn over to the endowment some of the large funds which I understand you are fully determined to devote to this purpose? If you could do so now, I would feel free to propose Throop at once to the Rockefeller Foundation as the California headquarters in the scheme.

At a dinner with Hale, Millikan, and Scherer on March 6, 1918, Fleming pledged $1 million unconditionally and promised to raise a second million.

Eager to seal the bargain, Millikan and Hale discussed the scheme informally with Vincent in New York several days later. They also told him of Fleming's offer to add $1 million to Throop's coffers. Vincent quickly pointed out to them, Hale reported to Scherer afterward, that "the Foundation was opposed to the provision of large endowments before a scheme was tested." Nevertheless, Hale had no intention of taking the pressure off Fleming. "So far as Throop is concerned," Hale told Scherer that March,

> I see no reason why we should not get Mr. Fleming to do all that he proposed, in order to secure the establishment of the Pacific Coast laboratory there. It is clear, if the decision is to be left to the Research Council, that Millikan, Noyes, and I, who are connected with Throop, could not favor it above Berkeley and Stanford unless it can make a stronger bid than they can.

A stronger bid on Throop's part, Hale went on to explain, would involve more laboratories, better equipment, as well as "other attractions for advanced students." But to turn the college into the best place for physics and chemistry on the West Coast, he warned Scherer, would require raising far more money than Fleming had just pledged. Indeed, Hale estimated Throop needed four million dollars, not one, to compete with the two other California institutions up north. Once the prime mover behind the effort to develop Throop into a fine engineering college, Hale now seemed intent on using Fleming's pledge to transform Throop into an institution that put the pure sciences first.

Whether Scherer liked it or not, Hale had not only raised the stakes but put him in charge of finding the needed millions.

Ambivalent about his own future as Armistice Day approached, Scherer was threatening to seek employment elsewhere if working conditions in Pasadena did not improve. He wanted a raise, among other things. Recruited by Hale in 1908 at a starting salary of $6,000, Scherer had apparently made a private deal with Fleming on the side. Inasmuch as Fleming paid his salary personally, Scherer apparently felt justified in making such an arrangement. Fleming, it seems, had promised Scherer in writing— "it was virtually a contract," Scherer later insisted—a bonus of $15,000, after he became school president. But Fleming destroyed the physical evidence, and Scherer never collected his bonus.

Nevertheless, Scherer spent much of the next decade patiently milking out of Fleming operating funds for the school. This took its toll. On February 22, 1918, Scherer wrote to Hale, in longhand, from a hotel in Chicago. "Before long I shall be fifty years old, and I do not intend to pass that milestone with a salary of $6,000 as my value scale, and no $15,000 contract to alleviate my embarrassment with the uplift of hope," he said with uncharacteristic candor and bitterness. Worse still, Fleming had openly bragged about keeping him on a short leash. "I must have something more tangible than promises for the endowment of Throop, if I am to stay there," Scherer warned. Reminding Hale that he was still considering the possibility of entering the California gubernatorial race, Scherer also recounted in his letter how Fleming had casually taken him aside months before and guaranteed "that he would 'cash in' without waiting for the war to end." "This is February and nothing more has happened that I know of," grumbled Scherer. His decade of cultivating Fleming's interest in the school had apparently come to very little.

The "Rockefeller scheme," as Hale liked to call it, marked the beginning of the end of Throop College of Technology. In the months that followed Millikan's first meeting with Rockefeller officials, the scheme itself went through numerous drafts and revisions. Scherer, describing himself as "profoundly interested" in the plan, urged Hale meanwhile to keep after Fleming. "I don't want . . . [him] to get 'cold' on his proposal," fretted Scherer. What Scherer had learned about Fleming over the years, Millikan was also beginning to learn. After one particularly grueling day in Washington, capped by a dinner with Hale and Fleming and a return to the office, where he worked past midnight getting a talk in shape for the following day, Millikan reported glumly to Greta, "We wished to get . . . [Fleming] interested in starting the ball rolling for the establishment of three Physics & Chemistry research laboratories. If he was persuaded he didn't turn over a ten million dollar check tonight, anyway."

Nor did anyone else. Having promised to raise the second million for Throop himself, Fleming had gone to see the southern California railroad tycoon Henry Huntington in New York, intending to ask him for five million. At the last moment, Hale persuaded Fleming to ask for one million instead. Huntington rebuffed him on the spot.

In the end, the request for funds to endow three research laboratories did not survive the protracted negotiations among the plan's principal backers—Hale, Millikan, and Noyes—and Rockefeller officials. Both sides agreed, instead, in spring 1919 to set up a postdoctoral fellowship program for research in physics and chemistry. The Rockefeller Foundation would pay the bills, while Hale's National Research Council, originally established as a wartime science advisory board, would select the fellows and administer the program.

The NRC, in fact, had survived the war unscathed. An executive order, which Hale drafted and President Woodrow Wilson signed in May 1918, transformed the organization into a permanent arm of the National Academy. The council's expanded charter, which now ranged from stimulating research in the basic sciences to promoting international scientific cooperation and gathering and disseminating scientific information, gave Hale and his indefatigable co-workers, Millikan and Noyes, the government connection they desired, without sacrificing the NRC's intrinsic autonomy. Hale had orchestrated the latter by reserving for the NRC the right to nominate the federal government's representatives. "Our relationship to the Government," he wrote to his wife, Evelina, shortly before Wilson's action, "is just what it ought to be . . . we could accomplish much less if we were a part of the regular governmental machine." The "we" referred to Hale, Millikan, and Noyes.

Not everyone agreed with the trio's relentless efforts in 1918 to circumscribe the role of the federal government in the development of science. If it is good for the country to have scientists, the director of General Electric's research department wrote to Hale that June, then it is equally important for democracy,

> to pay its own expenses in this line, just as in lower schooling. . . . I wish very much our people could see the importance of appreciating and backing their educational institutions because of the results they would obtain. Why should science, which is back of every one of our comforts, conveniences, and facilities today, be a matter the advancement of which, in the United States, is subject to the whims of philanthropists?

Hale's reply to General Electric's Willis Whitney was revealing. Standing by his decision not to make "advances to Congress," Hale turned the industrial manager's question around. "Do you think, for instance," he asked Whitney, "that any great good would result if all of our educational institutions were run by the Government and supported by tax payers?" Until such a time as elected officials really came to value research, Hale saw a continuing need for private funds and private institutions as partners in the development of science.

Against this background of money and politics, the challenge of rebuilding Throop gradually took on a life of its own. With or without backing from Rockefeller, Hale, Millikan, and Noyes now intended to push "the Throop end of the project" through. "I feel very sure that we will obtain all the money necessary to place the college where it belongs," an anxious Fleming wrote Hale in 1919. However, he pointed out that the income on the $1 million he had promised Throop would only replace, not increase, his current contributions. If Throop should expand in the direction favored by Arthur Noyes—that of pure science, with an emphasis on physics, chemistry, and mathematics, graduate education and research, national stature—it was going to take more than a couple of million dollars.

Efforts to get Fleming to endow a physics and chemistry research program at Throop limped along in 1919. That January, following a meeting with him in New York, Hale told Scherer to go easy on Fleming: "Millikan and Noyes were there, and everything went through beautifully and definitely. I put it up to him that he should give a definite promise of two million at once for chemistry and physics. . . . I advise one thing when he comes to Pasadena soon—don't press him for money. We carried the thing to the very limit, and he took it admirably, but it would do harm rather than good to press harder."

When Hale learned in February that Fleming could manage only $1 million as a gift, he told Scherer he would have to get more involved. "You should take active part," in raising the second million, Hale wired Scherer. But almost immediately Hale reversed himself. Citing hard times in Pasadena and elsewhere, he counseled Scherer to put the brakes on a public campaign for funds until the fall. "Talk this over with Fleming, and see if he does not agree," Hale recommended. Noyes, meanwhile, had drawn up a detailed battle plan for financing Throop's research program, and sent it to Scherer in Pasadena, calling for the president to solicit funds from the Rockefeller Foundation immediately. But it was Hale, not Scherer, who in March 1919 did the preliminary talking to Rockefeller officials about covering some of Throop's needed endowment.

By then, Scherer had become a part-time president. His health failed, and bouts of depression, which had plagued him when he taught English in Japan in the 1890s, returned. Referring in his letters to "a physical breakdown," Throop's president was permitted, under doctor's orders, to spend only an hour a day on school business. Still, he told Hale in March 1919 that he could make decisions and take the responsibility that went with being president, provided he paced himself. "Oddly enough," he added, "I look forward with pleasure to the Great Campaign, to which Mr. Fleming now gives his assent." The pleasure did not last long. Although he exulted in personally getting Fleming's pledge that spring—"I have in my pocket a sealed contract for $1,000,000, signed by A.H.F. . . . am still [on the] hunt for MORE"—by fall, the novelty had worn off.

"It may be a playful habit," he wrote Hale in September, "but I am getting extremely tired of having it 'rubbed in' that failure to obtain funds is due exclusively to the President, especially when he is forbidden to see the very group of men from whom contributions may most reasonably be expected." Troubled by his lack of success in raising money, Scherer drafted a letter of resignation in 1919, not long after one of his heroes, former President Theodore Roosevelt, died. While not intended for immediate use, the document provided, in his words, "psychic relief" from the burden of raising money for Throop. Hale, whose daughter had recently married Scherer's son, pleaded with him to stay on: "The whole question is so tremendously important, both to Throop and to you, that I hope you may withhold final judgment for some time."

During his last years in office, Scherer figured less and less in the decisions affecting Throop's future. Where is Throop headed? the college's business manager, Ned Barrett, asked the ailing president in 1919 as school officials mapped out an ambitious fundraising campaign. Keen to make more, not less, of the college's course work in engineering and economics, Barrett spoke for many on the campus when he lobbied for Throop "to continue to be a training school for engineers of various sorts as well as an institution for research and for the training of research men." Others, he admitted, dreamed of transforming the college into an institution emphasizing " 'pure' science and withdraw[ing] more and more from the technological side."

The events of the next few months, among them Arthur Noyes's momentous decision in December 1919 to leave MIT for good and accept a full-time position at Throop, sapped Scherer's endurance, rather than invigorating it. So did the news that month that the trustee Norman Bridge was prepared to put up the money for a physics building. Shortly after the school took the name Caltech, on February 10, 1920, Scherer resigned his office. He first tried his hand at writing movie scripts, and went on to become director of the nearby Southwest Museum. He later told Fleming, "The heavy responsibilities of my position have 'got on my nerves'—and *brain*—to a degree past all endurance." Scherer died in 1941.

Some months after Scherer's resignation, the mathematician E. B. Wilson at MIT wrote Noyes a letter, inquiring whether he was going to take the president's job at Throop. If not Noyes, then perhaps Millikan. "He should make you a good president . . . and be an inspiration to physics and science," Wilson added. The thought had already crossed Hale's and Noyes's minds.

FIVE

◆

Millikan and the Rise
of Physics

The creation of research men may not be the prime
function of all universities but it should certainly be
the prime function of some of them.

—ROBERT A. MILLIKAN,
1919

MILLIKAN was fifty-three when he came permanently to
Caltech in 1921. By then, he had piled up an impressive record as
the complete academic—teacher, writer, and researcher. Behind
him lay the measurement of the charge on the electron, the ver-
ification of Einstein's photoelectric equations, and the numerical
determination of Planck's constant. For these he won a Nobel
Prize in 1923.

But Robert Millikan had invented a future for himself long
before the invention of the Nobel Prize. The son of a Congre-
gational minister, Robert Millikan entered the world of physics
rather than metaphysics. In his memoirs, he described his strange
beginnings in it. Intending to major in classics, he had taken a
large amount of Greek at Oberlin and only one course in physics,
which he termed "a complete loss." But at the end of his soph-

omore year, Millikan's Greek professor nonetheless asked him to teach the elementary physics course the following year. When Millikan replied that he didn't know any physics, his professor's answer was "Anyone who can do well in my Greek can teach physics."

Millikan spent that summer working every problem in the physics textbook. In class that fall, he got all the students to do the same thing. By the time he had graduated from Oberlin with a master's degree, in 1893 (he received a bachelor's degree in 1891), Millikan had raised the teaching of physics to a fine art. Indeed, he went on to write a series of textbooks that were the mainstay of American physics instruction in the first half of the twentieth century.

Millikan, it seemed, could do anything. Being also a fine athlete, especially adept at boxing, he toyed with the idea of a career in physical education. But during his senior year at Oberlin came the unexpected news that he had won a graduate fellowship in physics to Columbia, and the notion of a career in physics appealed to him even more. Eighteen ninety-three was "the year," Millikan later wrote, "in which I decided to try to make of myself a real physicist if I could, a matter upon which I had some doubts for some years thereafter." He once calculated that it took him fifteen years to earn his physicist's stripes.

Robert Millikan's apprenticeship began at Columbia, where, as the only physics graduate student, he was well looked after by the staff. The physics department head, Ogden N. Rood ("a strange-appearing old man, with long stringy hair," is how Millikan remembered him), selected Millikan's first research problem—the polarization of light emitted by platinum and other incandescent materials. Besides getting a taste of experimental work, Millikan took advanced mechanics from R. S. Woodward, a mathematical physicist, and optics, electrical engineering, and electromagnetic theory from Michael Pupin, a physicist turned inventor. Courses in chemistry and mathematics rounded out Millikan's graduate education.

The best-known American physicist in Robert Millikan's day

was Albert A. Michelson, head of the physics department at the University of Chicago and director of research at the spanking-new Ryerson Physical Laboratory. Eager to make his acquaintance, Millikan spent the summer of 1894 at Chicago, where he enrolled in Michelson's graduate course in physics and also signed up for research work with him. A classy experimentalist, the inventor of the interferometer and other precision optical instruments, Michelson specialized in measuring the velocity of light and other fundamental constants of nature. When Millikan told him about the optical problem he was working on, Michelson "outlined precisely how he would suggest going at it." Then at the apex of his scientific career, Michelson made a deep and lasting impression on the budding laboratory scientist.

Returning to New York in the fall, Millikan finished up his polarization experiment, for which he received a doctor's degree in 1895. In that era, no American scientist's education was considered complete without a year or so of postgraduate study in Europe, preferably in Germany. Pushed into it by Pupin (who lent him $300, repayable at 7 percent interest), Millikan crossed the Atlantic that May, heading first for Jena, where he spent some time acclimating his ear to spoken German, and then setting out on a brisk two-month bicycle tour of the Continent, accompanied by a friend from Columbia. From Dresden, they pedaled to Venice, then on down to Rome, going as far south as Naples before heading back to Paris. "I expected a beautiful road from Florence to Rome," Robert wrote to his parents, proudly displaying a strong streak of American chauvinism. "I supposed, and suppose still, that the road was laid out by the Romans. If so, the glorious old Romans were glorious old idiots when it came to road building." In Paris, Millikan paused long enough to hear the great French mathematician Henri Poincaré lecture. He returned alone to Berlin in October 1895, completing the 700-mile route in one week flat.

In Berlin, Millikan picked up his books again. He registered for seventeen hours of course work, all in physics, including a lecture course in theoretical physics taught by Max Planck. The subject on everyone's lips in the laboratory in Berlin in 1895, he said later, was the nature of cathode rays (which the English physicist J. J. Thomson in 1897 correctly identified as subatomic particles, called electrons). Millikan was not impressed. In one of the few surviving letters written to his family from Berlin that fall, he stopped just short of saying the trip abroad had been a waste. "As far as I can see," he wrote in November, "the men here are no better than those whom I was under in Columbia altho [sic] there is a somewhat wider range of subjects offered." Even so, the spellbinding events of the next few months, including Wilhelm Roentgen's startling discovery of X rays in November 1895—followed by his exhibit of X-ray photographs of the human skeleton in January 1896, and Antoine Becquerel's discovery of radioactivity in uranium later that year—took the world by storm. All of them happened right under Millikan's nose.

In spring 1896, Millikan left Berlin and moved to Göttingen, where he worked in Walther Nernst's new physical chemistry laboratory and followed his lectures, besides taking Felix Klein's course in geometry and Woldemar Voigt's in thermodynamics. By then, he already had in his pocket an offer of an instructorship in physics from Oberlin. But when Michelson cabled him that summer with an offer of an assistantship at Chicago, Millikan didn't hesitate for a moment. He explained by return letter, "I have decided as I have, because I want opportunities to do research work."

The new university and the research-smitten Millikan got on well. "Those were great days on the banks of the Midway," one of his close friends from that time recalled, adding, "An unforgettable seething cauldron of intellectual curiosity on the raw campus with the drab backwash of the World's Fair all about and where wise men walked in the middle of the street after dark for fear of being stuck up." Millikan's chief fear was that he might lose "the gamble" that had originally brought him to Chicago. For while the university offered him the opportunity to do research work, it did so under conditions that would have crushed the determination of anyone less driven than Michelson's new recruit. Between 1896 and 1908, Millikan worked twelve hours a day, six hours on teaching and writing textbooks and six on research. By his own accounts, life had treated him reasonably well: he had a loving wife, a good marriage, two sons (a third son was born in 1913), a grand house on the edge of the campus, a tenured position at a prestigious university, and a steady stream of royalty checks.

Public success as an experimental physicist proved more elusive. "I knew," Millikan later wrote, "that I had not yet published results of outstanding importance, and certainly had not attained a position of much distinction as a research physicist." In 1908, he "kissed textbook writing good-bye" and started working intensively on what he called "My Oil-Drop Venture (e)." Millikan had two things to find out with that experiment—whether the electric charge came in quantized units and, if so, what the size of the unit was. The construction of apparatus to determine the value of e—the elementary unit of electrical charge—began in 1907, underwent a series of modifications and refinements by Millikan and his graduate students, and, in its final form, involved measuring the rate of fall of a single electrically charged oil drop falling in air under the forces of gravity and electricity.

Millikan's oil-drop experiment was an experimentalist's dream—theorists might think they knew that electricity was atomic in nature, but only the experimentalists could prove it beyond doubt. Although Millikan never said explicitly why he elected to work on this problem, he left little doubt that the fundamental importance of e in all fields of physics played a key role in his thinking. It may well be that Michelson's receipt of the Nobel Prize in 1907, for measuring the precise speed of light, also influenced Millikan. Indeed, in his memoirs, Millikan recalls talking to Michelson about how precise the oil-drop measurement would be, if the experiment worked. As Millikan tells the story, he said to Michelson, "I have in mind a method by which I can determine the value of the equally important universal physical constant, the electron, not to one part in 10,000 as you have done in the case of the speed of light, but to one part in 1,000 [i.e., to an accuracy of one-tenth of 1 per cent], or else I am no good." Michelson was famous for not paying attention to anyone else's research problems. But the oil-drop experiment "interested even him," Millikan once told his wife.

Millikan often singled out Michelson's style of physics, noting that observing him at work was like watching a master teacher perform. In Michelson's case, "the example . . . of always having on hand an experimental problem at which he worked every afternoon in his laboratory up to the time for his usual four-thirty tennis game at the Quadrangle Club was as good teaching by example . . . as could be found anywhere." Millikan, too, had an endless supply of research problems at hand. This was true all of his life—at the age of seventy-nine, six years prior to his death, he was still actively doing experimental physics.

In different ways, Millikan took what he learned from Michelson ("pure experimentalist") and about him ("he was an intense individualist") and used it to advantage in developing physics at Caltech when he went there permanently in 1921. Under Michelson, Chicago's physics department was strong on the experimental side and weak on theory—not surprising, perhaps, given Michelson's insistence, as Millikan put it, "on the place of very refined measurement in the progress of physics." It is no accident that, at Caltech in the 1920s and 1930s, Millikan's physics department excelled in experimental problems and left theoretical physics to fend for itself.

At Caltech, Millikan proved he could be just as single-minded with himself and others as his mentor Michelson had been. He drove a hard bargain with Caltech's presidential search committee, which consisted of the astronomer George Hale and the chemist Arthur Noyes. Hale and Noyes wanted to use Caltech to reshape the education of scientists; Millikan wanted, in his own words, "to put physics on the map" in southern California. To do that, he needed research funds. The three men came to an agreement. Hale and Noyes promised Millikan the lion's share of the school's financial resources and minimal administrative duties as head of the Institute. In return, Millikan agreed to come, as director of the Norman Bridge Laboratory of Physics and as chairman of the executive council of the Institute. (Strictly speaking, Millikan never served as president.)

The negotiations with Millikan did not affect Hale directly. Hale neither taught nor had graduate students at Caltech. As director of the Mount Wilson Observatory, he obtained his research funds from the Carnegie Institution of Washington. To meet Millikan's demands, Noyes, however, agreed to give up his promised share of Institute funds with which to expand the chemistry division. Physics, in any event, grew at the expense of chemistry and engineering during Millikan's tenure at Caltech.

Indeed, it grew by leaps and bounds. Millikan brought with him from Chicago a first-rate instrument maker by the name of Julius Pearson, as well as the research assistant Ira Bowen, who later received a Ph.D. in physics under Millikan. One of his former graduate students, Earnest C. Watson, was already in Pasadena, where he had gone in 1919, in part to monitor the day-to-day work of the physics department in Millikan's absence, in part to oversee the planning and construction of the new physics building, and in part to teach kinetic theory and thermodynamics, two brand-new courses.

Other Chicago recruits included W. T. Whitney, who taught advanced-undergraduate- and graduate-level courses in physical and laboratory optics; Russell Otis, who earned a Ph.D. in physics in 1924 from Caltech working on cosmic rays; and Ralph Smythe. Smythe got his Ph.D. at Chicago and joined the Caltech faculty in 1923, following a two-year stint in the Philippines. He remembers Millikan's telling him, "You'll ruin yourself. You'll spend two years out there and you're stuck." No such thing happened to Smythe. He applied for and received a National Research Council fellowship, which Millikan then persuaded him to use at Caltech.

Interested in the problem of separating isotopes—atoms that are chemically identical, but have different masses—Smythe later devised a method of separating the isotopes of iodine, rubidium, potassium, and other chemical elements in quantity, electromagnetically. He also began teaching advanced electricity and magnetism, which quickly gained a reputation among graduate students for being the toughest course in physics. At the end of his fellowship, Millikan saw to it that Smythe remained at Caltech.

William Houston, another NRC fellow at Caltech in the twenties, also went on to become a permanent member of Millikan's faculty. He, too, had fallen under the ebullient physicist's spell at Chicago. Houston's own explanation of how he happened to choose Caltech over Harvard in 1925 helps to explain Millikan's success in attracting outstanding researchers. "I went to Chicago in the summer of 1921," he later told an interviewer,

> and he [Millikan] was there for his last quarter, and I attended his class on quantum theory and saw him in some other connections and was tremendously stimulated by him. He had the knack—at least as it appeared to me and I think to many others—of making the subject of Physics sound like the most important thing in the world, although he wasn't always so very precise in his teaching. You had to go look up the things afterwards to understand them clearly, but he conveyed this idea of urgency and importance in a way which very few other people have—at least to me.

Although Houston doesn't say so, Millikan was awfully good at picking research men.

Houston himself is a good example. An experimental spectroscopist by training, he became interested in 1927 in applying the new quantum mechanical theories of Werner Heisenberg and Erwin Schrödinger to the fine structure of hydrogen and helium. By the end of that decade, Houston had bridged the gap between the experimental and the mathematically sophisticated physicist: in the laboratory, he carried out elegant experiments on the intensities in complex spectra; as a theoretical physicist, he grappled with the electron theory of metals. Houston taught introductory mathematical physics, an immensely popular course that had a wide following among the more theoretically minded chemists and engineers. Millikan had to stay one or two steps ahead of Stanford, MIT, and other schools bent on stealing Houston away.

It was Paul Epstein, Caltech's professor of theoretical physics, who tutored Millikan's prized experimentalist in quantum mechanics in the beginning. Epstein's professorial appointment came in 1921, right at the start of Millikan's career at Caltech. Epstein's story goes to the heart of Millikan's determination to make Caltech's Norman Bridge Laboratory the best-known address in the world of physics.

Born into a patrician Polish-Russian Jewish family, Epstein in 1901 entered the School of Physics and Mathematics at the Imperial University of Moscow, where he worked under P. N. Lebedev, an experimental physicist best known for measuring the pressure of light. Epstein earned a bachelor's degree in science in 1906 (the revolution of 1905 delayed his senior examinations by one year) and then enrolled as a graduate student. He was a laboratory instructor in physics, while conducting experimental research on the dielectric constant of gases. In 1909, he received a master's degree in physics and became the equivalent of an assistant professor. That December, at a scientific congress in Moscow, he met the theoretical physicist Paul Ehrenfest, who directed his attention to the West.

Having decided that he was not cut out to be an experimentalist ("my hands were not clever enough"), Epstein left Moscow in early 1910 for Munich. There he attended Arnold Sommerfeld's lectures on relativity; studied the theory of electromagnetic waves, particularly the theory of diffraction; and finally spent four years at Sommerfeld's Institute for Theoretical Physics. In 1914, Epstein received the Ph.D. in physics, with minors in mathematics and crystallography, from the University of Munich. During World War I, the Germans classified him as an enemy alien and interned him briefly. His interest in problems of quantum theory coincided

Amos Gager Throop (*left*) was the unlikely founder of a school that would one day evolve into one of the world's leading scientific institutions. Frontiersman, self-made businessman, and philanthropist, Throop poses here with his second son, George Throop, in Chicago in about 1860. A soldier in the Civil War with the Chicago Mercantile Battery, George Throop died in battle in 1864, at the age of twenty-four.

In 1910, the east side of Throop Polytechnic's first building had ready-made landscaping—a large and verdant orange grove bisected by a boardwalk. Throop Hall (dubbed Pasadena Hall for its first ten years) was the site of classes, then of administrative offices, until it was torn down, in 1973.

The first school west of Chicago to teach manual skills, Throop Polytechnic Institute taught—as its mandate proclaimed—"those things that train the hand and the brain for the best work of life." The modeling class in this turn-of-the-century photo included students of all ages. Throop also offered training in piano, voice, painting, and drawing.

An aerial view of the Caltech campus in 1922, when the school consisted of twenty-two acres of land and four permanent buildings: Throop Hall, for engineering; *left*, Gates, for chemistry; *right*, Bridge, for physics; and, *foreground*, Culbertson Auditorium. *Lower right-hand corner*, Pasadena's original Rose Bowl.

Robert A. Millikan and his wife, Greta, setting out for the Harz Mountains, Germany in 1912. Left behind was their six-year-old son, Glenn.

TED DUCEY
EXCAVATING CO.
CO. 1044
INSURED WITH
ROY JORDAN INC.
571 E. GREEN PASADENA CO. 5348
ROOSEVELT for KING
JESUS SAVES
BUT MILLIKAN GETS CREDIT

Millikan inspired admiration and respect across the country as the foremost scientific spokesman of his day, but the attitude on his own campus tended to be somewhat less reverent, as this specimen of graffiti from 1937 shows. The occasion was the construction of the second unit of Caltech's Kerckhoff Laboratories of the Biological Sciences.

The solar astronomer George Ellery Hale played a key role in transforming Throop Polytechnic Institute into the California Institute of Technology. Hale in 1904 also founded the Mount Wilson Observatory, in Pasadena, California, with funds provided by the Carnegie Institution, and served as observatory director until 1923.

The lumber magnate Arthur Henry Fleming joined Throop Institute's board of trustees in 1903 and went on to become one of Caltech's principal benefactors. He and his daughter donated the property that became the nucleus of the Caltech campus, and Fleming subsequently turned the bulk of his fortune (in excess of $4 million) over to the Institute in a successful bid to lure Millikan permanently to Pasadena. Unfortunately, the Great Depression wiped out most of Caltech's Fleming Trust.

James A. B. Scherer, president of Newberry College, in South Carolina, was hired by Hale in 1908 to be the first president of the "new Throop." He served until 1920.

To practice for the battlefields of World War I, the Student Army Training Corps dug trenches all over Throop's campus and attacked every available stronghold.

The only known photograph ever taken of Caltech's ruling triumvirate on the campus: *right*, Robert A. Millikan; *center*, George Ellery Hale; and Arthur A. Noyes. A campus wag christened them "tinker, thinker, and stinker."

Carl Anderson discovered the positron—the first empirical evidence for the existence of antimatter—in 1932, by means of this magnet cloud chamber. Housed in the Guggenheim Aeronautical Laboratory, the apparatus was in service for seventeen years. For his discovery, Anderson shared with V. F. Hess the Nobel Prize in physics in 1936.

The versatile Richard Chace Tolman already held full professorships in physical chemistry and mathematical physics when he became a leading cosmologist in the 1920s. During World War II, he served as scientific adviser to General Groves on the Manhattan Project.

Caltech physicists and mathematicians in 1932. *Front row, from left*: J. Robert Oppenheimer, Harry Bateman, Richard C. Tolman, William V. Houston, Robert A. Millikan, Albert Einstein, Paul S. Epstein, Fritz Zwicky, and Earnest C. Watson.

Albert Einstein's visits to the campus in 1931, 1932, and 1933 capped Millikan's campaign to make Caltech one of the physics capitals of the world.

Linus Carl Pauling entered Caltech as a graduate student in 1922, completing his Ph.D. in chemistry in 1925. Then he sailed for Europe, where he passed the next nineteen months in the company of its finest theoretical physicists. While abroad, he was appointed assistant professor of theoretical chemistry at Caltech. By the time Pauling returned to Pasadena in 1927, his quest to formulate a quantum theory of the chemical bond had begun. Pauling received the Nobel Prize in chemistry in 1954 for his research on the nature of the chemical bond and its application to the understanding of the structure of complex substances.

John Merriam, professor of paleontology at the University of California at Berkeley, was tapped by George Ellery Hale to be president of
the Carnegie Institution of Washington in 1920. He broke with Hale
in the 1930s over the issue of establishing a Caltech astronomical observatory on Palomar Mountain.

with the publication in 1916 of Sommerfeld's paper on the fine structure of atomic hydrogen.

Epstein wrote a series of important papers on quantum theory and its applications. In his classic paper in 1916 on the theory of the Stark effect, the splitting of the spectral lines of a hydrogen atom by a strong electric field, he worked out the quantization rules in an invariant form and then used them to calculate the splitting of the hydrogen lines. The splitting effect, first observed by Johannes Stark in 1913, could not be explained along classical lines. Epstein's demonstration that Niels Bohr's description of the hydrogen atom could be used to solve the problem made his reputation as a theoretical physicist. The match between his theoretical predictions and Stark's data furnished striking support for the Rutherford-Bohr atomic theory. It also brought praise from Sommerfeld and from another theoretical physicist on the Berlin faculty, Albert Einstein, then holding a temporary teaching appointment at Zurich. By 1921, Epstein was an assistant to the Nobel laureate Hendrik Lorentz at Leiden, but bent on finding a permanent position in America.

Millikan met Epstein for the first time in Leiden that April, and they discussed the possibility of Epstein's teaching at Caltech. At the time, Hale and Noyes were pressing Millikan to break completely with Chicago. Millikan was in Europe to attend an international physics conference in Brussels. From there, he took the train to Leiden, where he gave a physics colloquium at the university, dined with Heike Kamerlingh Onnes, the distinguished Dutch low-temperature physicist, and spent a memorable evening at Paul Ehrenfest's home. "I found myself," Millikan wrote back home to his wife in Chicago the next morning, "in an interesting and unique sort of an atmosphere." Ehrenfest, he went on to say,

> is I suppose a Polish or Hungarian Jew [actually, Austrian] with a very short stocky figure, broad shoulders and absolutely no neck. His suavity and ingratiating manner are a bit Hebraic, (unfortu-

nately) and to be fair perhaps I ought to say too that his genial open-mindedness, extraordinarily quick perception and air of universal interest and inquiry are also characteristic of his race.

His personal opinions aside, Millikan seems to have made the most of his brief stay in Holland. He secured Lorentz's visits to Caltech in 1922 and 1927, Ehrenfest's in 1924, and Epstein's in 1921.

In June 1921, Millikan drafted a letter to Epstein inviting him to come to Caltech for a year, starting that September. This letter, dated June 3, was never sent. Instead, Millikan wrote on July 7 to C. G. Darwin (grandson of Charles Darwin), at Cambridge University, extending the same invitation, but for the academic year 1922–23—which Darwin accepted. This left Millikan, who by now had decided to sever his Chicago ties, with no one to teach the "stiff graduate courses" in theoretical physics at Caltech that fall. Millikan considered Epstein's case again. "I am still hesitating about Epstein," he wrote Hale, "but will certainly write him at once if Darwin can't come. Do you still think we want to get Epstein anyway? Even though a Jew!" Once again, Millikan wasn't shy about discussing the subject of Jews among friends. Why did he raise the question now with Hale? The simplest explanation is that Hale and Noyes had voiced their objections when Millikan first brought Epstein's name up, and persuaded him not to send the original letter. Indeed, Hale continued to counsel Millikan to do nothing about Epstein before coming to Pasadena in October.

Instead, Millikan sent Epstein a letter on August 5, inviting him to come for one year. In explaining his actions to Hale, Millikan wrote, "We can keep Epstein busy this year certainly in training our advanced group in mathematical physics, in getting them more familiar with German literature in this line, and in helping on the particular research problems which you and I are interested in." Indeed, Epstein later complained that once he got to Caltech, he never had enough time to do his own research.

Epstein's boat docked in New York on September 16, 1921, and from there he boarded a train for the trip west. A telegram from Millikan, then on the West Coast, reached Epstein when the train stopped in Chicago. He was to wait there until Millikan arrived. Epstein waited. When the two men met face-to-face three days later, Millikan got right to the point. Did Epstein have his lectures prepared? Yes, the theorist replied. If he thought the question peculiar, he said nothing. Millikan himself had no further questions. Why did Millikan go through this exercise? Perhaps he wanted, or needed, to reassure himself that he had done the right thing. In any event, the two men shook hands and Epstein continued to Pasadena.

During his first year at Caltech, Epstein wrote to the physical chemist G. N. Lewis at UC Berkeley, inquiring about a job for the next academic year. Lewis wrote back, according to Epstein, that "a good neighbor policy didn't permit them to take me." Epstein added, "With that, I went to Millikan, and he said that it was a permanent appointment. . . . I was on the faculty as Professor of Theoretical Physics. I had no contract, no tenure, no nothing—just the word of Millikan."

Paul Epstein came to play a significant role in the foundation of Caltech's physics division, introducing and teaching virtually all the theoretical physics courses in the early years. With Robert Millikan, he organized and ran the weekly physics research seminar, which became a Thursday afternoon tradition. Actually, physics had three colloquia weekly, including the Astronomy and Physics Club, which was a cooperative venture of Bridge Laboratory and the Mount Wilson Observatory.

Millikan initiated a visiting-scholars program shortly after his arrival in Pasadena. A. A. Michelson, the first scientist to be invited, was appointed a research associate in 1920. The list of other scientists who accepted Millikan's invitation represented the cream of European physics, including Niels Bohr, Max Born, Paul Dirac, Peter Debye, James Franck, Max von Laue, Paul Langevin, Erwin Schrödinger, and Werner Heisenberg. Albert

Einstein's visits to the campus in 1931, 1932, and 1933 capped Millikan's campaign to make Caltech one of the physics capitals of the world. If nothing else, Einstein's visits showed dramatically that the Caltech that Hale, Noyes, and Millikan had set out to build in the twenties had come of age in the thirties.

Robert Millikan liked to say that even if Einstein had never published a word on relativity, his other theoretical researches would have won him an enduring place in the history of ideas. But it was, in fact, the cosmological implications growing out of the theory of relativity that brought Einstein to the campus. Einstein had spent eight years transforming his ideas on the electrodynamics of moving bodies into the more comprehensive general theory of relativity. In 1907, he began toying with the problem of incorporating gravitation into the special theory, and in 1916 he published the fundamental paper on the theory of general relativity—which made a number of predictions. In 1919, two astronomical expeditions independently confirmed one of them— the bending of a ray of light in the vicinity of the sun during an eclipse. This dramatic news—which made the front page of the *New York Times*—almost single-handedly turned Einstein, the theoretical physicist, into a twentieth century folk hero and a show in himself.

Will Rogers described this show when he said, just after Einstein returned to Berlin in March of 1931,

> The radios, the banquet tables and the weeklies will never be the same. He came here for a rest and seclusion. He ate with everybody, talked with everybody, posed for everybody that had any film left, attended every luncheon, every dinner, every movie opening, every marriage and two-thirds of the divorces. In fact, he made himself such a good fellow that nobody had the nerve to ask him what his theory was.

"What his theory was" was very much the point of his three visits to Caltech, however.

Richard Tolman was Caltech's relativity expert at the time.

Tolman's scientific interests were varied, but the main thrust of his work at the Institute included statistical mechanics, relativistic thermodynamics, and cosmology. He had come to Caltech in 1922 as professor of physical chemistry and mathematical physics (and dean of the graduate school), through the efforts of Noyes. Seven years later, the Mount Wilson astronomer Edwin Hubble had made the discovery that redshifts of the spectra of galaxies are proportional to their distances from the observer—a finding that led inescapably to the conclusion that the universe was expanding. Spurred on by Hubble's discovery, Tolman undertook in the 1930s a series of studies on the application of the general theory of relativity to the overall structure and evolution of the universe.

Hubble's discovery challenged Einstein's cosmological picture of a static universe. The big question at Caltech in 1931 was whether Einstein would give up his cosmological constant and accept the idea of an expanding universe. By day, Einstein discussed his theory and its interpretation at length with Tolman, Hubble, and the other scientists on the campus. By night, Einstein filled his travel diary with his personal impressions:

> January 2, 1931: at Institute (with) [Theodore von] Kármán, Epstein, and colleagues. . . . January 3: Work at Institute. Doubt about correctness of Tolman's work of cosmological problem, but Tolman turned out to be right. . . . January 7: It is very interesting here. Last night with Millikan, who plays here the role of God. . . . Today astronomical colloquia, rotation of the sun, by [Charles E.] St. John. Very sympathetic tone. I have found the probable cause of the variability of the sun's rotation in the circulatory movement on the [surface]. . . . Today I lecture about a thought-experiment in the theoretical physics colloquium. Yesterday there was a physics colloquium on the effect of the magnetic field during crystallisation of the properties of bismuth crystals.

While in Pasadena, Einstein remained silent on his cosmological problems. Five months later, though, Einstein wrote to Millikan from Berlin that "further thought regarding Hubbel's [*sic*] observations have proved that the phenomena adapts itself [*sic*] very well to the theory of relativity." Within a matter of months, Einstein publicly adopted the expanding-universe model.

Tolman was Caltech's unofficial toastmaster. His opening remarks at one of the many dinners in Einstein's honor show how he earned that distinction and illuminate life at Millikan's Caltech of that era:

Fellow Scientists: First of all I should like to explain to you the reason why I happen to be toastmaster this evening. Three weeks ago today in the late afternoon I was strolling back and forth on the Institute Campus, buried in meditation, trying to find a solution for the terrible problem of the increase in entropy that appears to be taking place everywhere throughout the universe. Just at the moment when it seemed as if I were about to get a solution for the problem, my walk was suddenly interrupted by Dr. Millikan.

"Tolman," he said. "Yes, Professor Millikan," I replied—Dr. Millikan is an older man than I am and he always speaks to me in that informal way. He just calls me Tolman. But I am a younger man than he is, so I always reply, "Yes, Sir." "Yes, Professor Millikan."

"Tolman," he said, "I think it would be a good plan if we had a dinner at which the members of the scientific staff of the Institute and neighboring institutions could meet Professor Einstein." "Dr. Millikan," I replied, "I think that would be very fine for the staff members but pretty hard on Dr. Einstein. I am sure that in the course of his life he has had to attend so many dinners in his honour that he never wants to look another filet mignon in the face. I therefore recommend strongly *against* such a dinner."

Two weeks ago today, I was again strolling back and forth on the campus, and had again nearly reached a solution of the problem of entropy, and was again interrupted by Dr. Millikan. "Tolman," he said, "I have been thinking about *your* suggestion that we ought to have a staff dinner in honour of Dr. Einstein, and I believe we ought to have a number of speeches at the dinner by staff members." "Dr. Millikan," I replied, "I think that would be fine for the speakers but very hard on Dr. Einstein and the other listeners. I therefore recommend strongly *against* any speeches."

One week ago today, I was again strolling back and forth on the campus, and had again nearly reached a solution of the problem of entropy, and was again interrupted by Dr. Millikan. "Tolman," he said, "I have been thinking about *your* suggestion that we ought to have speeches at the staff dinner in honour of Dr. Einstein. Here is the list of speakers and I have decided to appoint you the toast-master."

That, Fellow Scientists, is the reason why I am toastmaster to-night and the reason why the problem of the entropy of the universe still remains unsolved.

To his relief, Einstein found that his subsequent visits to Caltech, in 1932 and 1933, attracted much less public attention. During his third visit, Einstein sidestepped as long as possible the question of whether conditions in Germany might prevent his return there. After the January 30 announcement that Hitler had become chancellor of Germany, the question could no longer be evaded. Scheduled to leave Pasadena at the end of February, he postponed his trip for a few weeks and then went to Belgium for several months instead of to Berlin.

In the fall of 1933, Albert Einstein returned to the United States as an émigré, landing, like so many before him, in New York City. From there he went the short distance to Princeton, New Jersey, to become a charter member of Abraham Flexner's new Institute for Advanced Study and, eventually, a citizen of the United States. Why did Einstein go to Princeton and not Pasadena?

The record starts with a letter from Noyes to Hale dated October 9, 1931. In the spring of 1931, Arthur Fleming, chairman of the board, Caltech's chief benefactor, and a man of increasingly erratic behavior, had offered Einstein a great deal of money— $5,000 annually in addition to $15,000 for a ten-week stay in Pasadena, plus an annuity for Mrs. Einstein—if Einstein would agree to a visit every year. "There was no outbreak against Fleming because of his commitment," Noyes reported, but a very strong undercurrent of feeling. Indeed, the one good feature in the matter is that it has, I believe, convinced every trustee that the present situation is an impossible one [a reference to Fleming's mental instability]. However, all the business men (even more than we academic people) seemed to feel that Fleming has absolutely committed us, and that we could not possibly back out from the arrangement for the *present year*, even though there was no money in sight. Fleming read *not* his own letters to Einstein, but two replies he had received from Einstein, the second one written in July stating that he (Einstein) definitely accepted the permanent appointment. . . .

Fleming, however, refused to reveal the contents of his reply to Einstein. "It seems *probable*," Noyes added, "that Fleming told Einstein that the arrangement was definite for this year, but that the permanent plan must await action by the trustees."

The trustees finally agreed on October 1 that if indeed Einstein had been offered $20,000, they would have to pay it. The board secretary, Barrett, cabled Millikan that same day (he was traveling in Europe at the time) to go see Einstein. "If Einstein does not feel we have made such commitment," the cable said, "Board voted that you offer fifteen thousand or thereabouts" for the coming year, leaving the "question of a continuing connection with Institute" to be discussed later on.

Millikan went immediately to Berlin to see Einstein. Einstein responded to Millikan's visit with a letter to Fleming, dated October 19, 1931.

Most esteemed Mr. Fleming! Dr. Millikan has presented your new offer to us. We have decided, thereupon, to remain in Europe this winter. We base our decision on the fact that I am opposed to accept the invitation in the face of the current financial negotiations. . . . I have decided to spend the winter in the southern European sun to recuperate from the exhaustive negotiations and to gather renewed strength for future [negotiations].

However, at this point, Elsa, Einstein's wife, apparently intervened. She wanted to spend the winter in southern California.

She wrote, "Dear Professor Millikan: I am returning the enclosed contract signed by my husband and will presumably embark on Nov. 28." The minutes of the next board meeting indicate the deal that was struck: "By the authority given me by the Board of Trustees of the California Institute of Technology, I herewith extend to Professor Albert Einstein an invitation to join the staff for the winter quarter of the academic year 1931–32. . . . It is understood that the compensation which the Institute will make to Prof. Einstein for his service is to be $7,000." The penny-pinching Millikan had saved Caltech a tidy sum of money, and coincidentally lost a permanent faculty member. When Einstein finally had to leave Germany, he wound up going to Princeton.

Even so, Caltech had Millikan, and next to Einstein, Millikan was during his lifetime undoubtedly America's most public figure in science. When Millikan spoke, the country listened.

How well informed the public was about Caltech in the thirties is indicated by a story about Millikan related by Carl Anderson, the first member of Millikan's physics staff to become a Nobel Prize winner. Anderson, who discovered the antielectron, or positron, in 1932 in the course of his cosmic-ray researches, spent four years as a Caltech physics undergraduate, stayed on as a graduate student of Millikan's, and subsequently earned his Ph.D. at the Institute. Anderson was on a train going to a physics meeting, when he fell into a conversation with another passenger in the club car. The fellow asked what he did. Anderson said he was a professor. He asked, "Where?" and Anderson said, "At Caltech." "Oh, is that part of UCLA or is it part of USC or what is Caltech?" "No, it's an independent college; it has nothing to do with SC or UCLA," replied Anderson. Then Millikan's name was mentioned and the fellow exclaimed, "Oh, you mean Millikan's school!"

Physics was king from the very beginning. It had more students, more faculty, and more money than other departments had. In 1927, Caltech had 500 undergraduates (the trustees, in 1921, had limited the freshman class to 160 students), 60 graduate students in physics, 20 in chemistry, and 4 in electrical engineering. It had attracted some 25 postdoctoral research fellows in physics, compared to 18 at Princeton, 16 at Harvard, and 14 at Chicago. No institution, Millikan boasted, had a larger number of NRC fellows in residence.

Caltech physicists were also prolific. More than 20 percent of the 348 major contributions to the physics literature during the period from July 1925 to August 1926 came from Millikan's own physics department. Indeed, Caltech topped the list of fourteen institutions in the country with 66 physics publications during this same period; the runners-up included Harvard with 59, Chicago with 39, and Princeton with 34. Staff members taught one course, seldom more than one, at any given time. Heavy teaching loads and research did not mix, according to Millikan. "I do not believe at all in the attitude that Millikan took when he was here last spring to the effect that we could not expect any research from men who had to teach more than four hours per week," a peeved MIT faculty member complained to Hale in 1919. "We have in physics today, a staff of thirty-one men at the Institute and if I should assign four hours' teaching per week, I should quadruple the staff."

Millikan himself juggled administrative, teaching, and research duties at Caltech. He taught a first-year graduate course on atomic physics called "Electron Theory," laughingly referred to among the students as "Meet the Chief." In class, Millikan was famous for reminiscing about the oil-drop experiment. "The real purpose of the course," Victor Neher later told an interviewer, "was to learn to know the graduate students. . . . And on the basis of that, he would try to assign the students research problems." In fact, Millikan had an uncanny knack for choosing significant physics problems, and he personally assigned them to graduate students. In Millikan's absence in 1926, Earnest Watson told the undergraduate Carl Anderson to work with a new NRC fellow by the name of Lee DuBridge. When Millikan returned three weeks later, he called Anderson into his office and reassigned him to work with D. H. Loughridge. Several weeks later, Loughridge received his Ph.D. and departed, leaving Anderson with a roomful of apparatus, including a cloud chamber, a handy piece of equipment for detecting positrons, as it later turned out.

The number of Caltech physics dissertations climbed from 4 in 1924 to more than 135 by 1940; Millikan supervised about one-third of these Ph.D.'s. He was a familiar sight around the physics laboratory, especially late at night, coming in after social functions, in evening clothes, and talking to students and co-workers about their research work. As Ira Bowen later recalled, his graduate student collaboration with Millikan on the extreme ultraviolet spectrum took place mainly after dinner. "In fact, often he [Millikan] didn't quite know what I was doing until I'd go around and say: 'I've got an article. How about coming around tonight?' He was always busy in the daytime, so he'd appear about nine o'clock in the office and we'd work until midnight writing the article. He always wrote the article . . . we worked together and published jointly."

The Caltech students also became Caltech scientists. Bowen was one of Caltech's first Ph.D.'s in physics, earning his degree in 1926. Millikan hired him on the spot as a member of the physics staff. By 1930, his four physics colleagues—DuMond, Lauritsen, Anderson, and Neher—were all Caltech Ph.D.'s.

Jesse DuMond's association with Caltech predated Millikan's. He had entered Throop in 1912, graduating with a B.S. in electrical engineering in 1916. In 1921, DuMond returned to Caltech as a graduate student, obtaining a Ph.D. in 1929. In DuMond, Millikan found a relentless, frugal researcher, a producer of ingenious equipment, and an idealist, a man fiercely loyal to the Institute. He voluntarily resigned a graduate teaching fellowship in 1924 to devote more time to research. Financially independent, he spent nine years (1929–38) as an unpaid research fellow, before finally being on the payroll as an associate professor.

DuMond's Ph.D. thesis dealt with the Compton scattering of X rays by atoms and with the breadth and structure of the shifted line. In his derivation of the shift in 1922, Arthur Compton assumed an elastic collision between a photon of light and an electron at rest. But if the atomic electrons had an initial momentum, DuMond reasoned, it could explain the observed broadening of the shifted line shapes. Armed with this tantalizing hypothesis, he undertook in 1925 a detailed spectral study of the Compton line, using the metals aluminum and beryllium as the scattering bodies.

The velocity distribution of the electrons doing the scattering, he argued, impressed a Doppler effect on the scattered beam, resulting in a broad, diffuse shifted line. Now considered one of the classical experiments of atomic physics, DuMond's measurements provided the first direct experimental evidence of the momentum distribution of electrons in atoms, as predicted by wave mechanics.

Like Noyes, Millikan systematically tapped his own students from the 1920s on. By 1930, Caltech was ranked as the leading producer of important physics papers in the country. "This productivity" reflected the efforts of the younger faculty, explained Caltech's leaders, who added that every younger member of the physics department had been "essentially grown on the spot."

Millikan believed that the modern world was basically a scientific invention, that science was the mainspring of the twentieth century, and that America's future rested on the promoting of basic science and its applications. Caltech, in Millikan's view, existed to provide America's scientific leadership.

He also knew just what it would take to train that leadership. "If a man does not learn his physics, chemistry, and mathematics in college, he never learns it," he told a Caltech audience in 1920. True to his word, Millikan instituted the basic undergraduate requirement of two years of physics, two years of mathematics, and one year of chemistry. That same curriculum is in force today at the California Institute of Technology.

SIX

•

Biological Work

From the standpoint of physical environment, I should
say that many points in Southern California are almost
ideal for scientific pursuits. From the standpoint of
social and intellectual environment, on the other hand,
the situation is quite different, at least in the extreme
south. We are wholly lacking in stars of the first magnitude, and even ones of the second magnitude are
extremely scarce. One must except, of course, the Pasadena group, but they are not biologists.

—FRANCIS C. SUMNER,
1923

THROOP changed its name to Caltech in 1920. In the next
ten years, more than one million people flocked to southern California, mostly settling in Los Angeles. The population explosion
created boom times for bankers and businessmen. Building speculation flourished; interest rates soared; and new real estate sub-
divisions dotted the landscape. Economically, the region
prospered as never before. Tourism, motion pictures, and oil put

millions of dollars into the pockets of those who had come west earlier. "In the bonanza years from 1920 to 1930," wrote Carey McWilliams, the American social critic and expert on twentieth-century California history, "Los Angeles had all the giddiness, the parvenu showiness, and the crazy prosperity of a gold-rush town."

Behind the wall of prosperity was a community of new-comers—"refugees from America," one writer called them—struggling to reconcile its diverse elements. The region boasted large numbers of wealthy, well-educated people, some of them scientifically inclined, like Charles Frederick Holder, a well-known naturalist, who had once taught zoology at Throop. And, as always, southern California attracted the fringe, a succession of quack doctors, faith healers, mystics, and messiahs. "We have," wrote a psychologist, Francis Sumner, to a colleague at Yale in 1923, "a large element of intellectual freaks—Christian Scientists, Theosophists, 'New Thoughters,' Anti-vivisectionists, and the like—who are positively hostile to science, and are an actual menace to the freedom of research." Under these conditions, advocates of nontraditional medicine did a brisk business. One thing is certain: there were more osteopaths, chiropractors, and natu-ropaths in Los Angeles in the 1920s than in all other California cities combined.

Whatever its deficits, southern California had students aplenty. By the midtwenties, the region boasted more than 12,000 students in private colleges and universities. Another 6,500 college students were attending the new southern branch of the University of California (now UCLA), which was still restricted to undergraduate education. While the region's population boom swelled the ranks of college-bound students, no other southland school, private or public, could match Caltech's strength—if the fields in question were physics, chemistry, or engineering. For despite its name, in the early twenties, Caltech was essentially an undergraduate and graduate school in the physical sciences. Indeed, until 1925, the institution offered graduate work leading to the doctorate only in physics, chemistry, and engineering. Caltech might not be as well rounded as other institutions of higher education, but as one knowledgeable official at the Rockefeller Foundation at the time remarked, "These graduate departments represent unoccupied fields in Southern California." In establishing Caltech, Hale, Millikan, and Noyes had led from their strengths—and these did not include the life sciences.

Nevertheless, if biology at Caltech had a godfather, that man was the chemist Arthur Noyes. It was Noyes who championed biology's cause, kept after Millikan to do something about it, and demanded room in the curriculum for it. Noyes had basically three arguments. "Aside from its importance as a basis of medicine," he wrote Millikan on one occasion, "biology must be established at the Institute, because it is an essential foundation for sanitary engineering and municipal hygiene, and because it is a highly important cultural study in the all-round training of a broad type of engineer, chemist or other scientific man." Seeking to capitalize on Caltech's existing strength, Noyes urged Millikan to start out with a research program in biochemistry and biophysics.

Noyes jumped into biochemistry in 1922, one year after the discovery of insulin. A Los Angeles physician by the name of Bernard Smith had guaranteed Caltech funds to pay for research in improving the technique of extracting insulin from the pancreas of animals. Noyes wasted no time purchasing equipment and finding space in the basement of the Gates Chemical Laboratory for the new research. Humanitarian reasons guided his decision, along with the desire to better the Institute's standing with two of the country's leading philanthropic organizations, the Carnegie Corporation of New York and the Rockefeller Foundation. "It would probably make much easier future development in these directions," Noyes reported that December to George Ellery Hale, Caltech's master planner, "if we can show . . . that we are already doing effective work in a small way in this field." In the space of a few months, Noyes reported progress in the preparation of insulin: Caltech's supply, he told Hale, was "twice as active a preparation as any obtained elsewhere," and they had a surplus, a research boon.

In summer 1923, Wickliffe Rose, president of the Rockefeller Foundation's General Education Board, toured the West Coast, visiting Caltech, among other schools. Hale was traveling abroad, but Rose talked with Millikan and Noyes about a possible medical school in Pasadena. Rose was followed by Henry S. Pritchett of the Carnegie Corporation. The chairman of Caltech's trustees, Arthur Fleming, quickly arranged a small dinner party on August 16 in Pritchett's honor. Other trustees on the guest list included the banker Henry Robinson, the newspaper publisher Harry Chandler, and the lawyer Henry O'Melveny, all prominent southern California tycoons. After dinner, Millikan spoke of Rose's interest in seeing a first-rate medical school established on the West Coast. Caltech's trustees listened politely, but they did not reach for their checkbooks.

Pritchett did. He took Noyes aside the next day and quietly guaranteed him $10,000 in research funds from the Carnegie Corporation to continue Caltech's biochemical work on insulin. As for establishing a large research-oriented medical school in Pasadena, Pritchett seemed receptive to that too. To hear Millikan tell it, both Rose and Pritchett agreed that "if anything is done in Southern California in the field of biochemistry, biophysics, and medical education, it must be done in immediate contact with the present work of the Institute." According to Rose, continued Millikan, the Rockefeller board, would "not be interested in any medical plan in Southern California which is farther away than across the street at most from the Institute."

Millikan was prepared to make space on the Caltech campus for medicine immediately, if it would bring about a successful launching of biochemistry and biophysics. "Your imagination can work on this plan just as well as mine," he wrote to Hale, then traveling around Europe. Enthusiasm aside, Millikan's plan had a practical side to it, for he would have liked nothing better than to forestall any local institutions from competing directly with Caltech for funds to develop the sciences, from astronomy to zoology. All the sciences belonged at Caltech, Millikan had once told Hale.

While Millikan indulged in these dreams, Hale had night-mares of Caltech's bankrupting itself. Upon his return to America in October 1923, Hale huddled with Wickliffe Rose in the Rock-efeller Foundation's headquarters in lower Manhattan. He felt personally responsible for Millikan's and Noyes's being at Caltech to begin with, Hale told Rose, and he was certainly willing, as he put it, "to back them up in all sound plans." A medical school, however, was an "enormous responsibility" that Caltech was nei-ther administratively nor financially prepared for. To Hale, it made more sense to bring biochemistry and, ultimately, biology— sciences basic to medicine, in any case—up to the same level as that of physics and chemistry at the Institute. What did Rose think? Rose replied that he could hardly claim to be an expert in such matters. His recent trip up and down the Pacific Coast was only his first. San Francisco, in fact, had impressed him as the best place to establish a medical school. "Los Angeles," he de-clared, "has nothing to offer in the direction of the development of a medical school." Pasadena, in short, was simply a better bet in southern California than Los Angeles was.

"The sound thing to do" at Caltech, they agreed, resuming the discussion over lunch at the University Club the next day, was to concentrate on building a "solid foundation" by devel-oping the core sciences. Would the Rockefeller's General Edu-cation Board be willing to finance a portion of that development? Hale asked. The board would welcome any Caltech proposal, Rose assured him.

In the meantime, Caltech's insulin project had turned into research on the substance itself. In fall 1923, the chemistry staff abandoned the laborious and time-consuming preparation of in-sulin when Eli Lilly and Company, the pharmaceutical house in Indianapolis, offered to supply the lab with sufficient quantities of the hormone for research purposes at a very modest price. Using his Carnegie grant, Noyes purchased seventy-five grams for $1,000, one-quarter of the cost a year earlier. By then, Noyes had also lined up several graduate students in chemistry to work on the insulin problem.

Noyes's interest in insulin lay in two directions. One promising line of research, the chemist wrote in his first progress report to the Carnegie Corporation, involved "attempts to develop an accurate *in vitro* method of determining the content of active insulin in insulin preparations." In fact, Noyes and his students did carry out several experiments on the effect of insulin on milk cultures, and they concluded that "the rate of lactic fermentation of glucose by bacteria" increased dramatically in the presence of insulin. However, Noyes was not a skilled biochemist, and these experiments led nowhere. The second line of research, carried out by his graduate students Gordon A. Alles and Albert L. Raymond, centered on isolating and purifying insulin—a formidable problem even for a highly skilled biochemist. Indeed, laboratories around the world were working on the new substance, and competition among them was fierce.

Faced with the prospect of coming up empty-handed in the laboratory and of not having his insulin research grant renewed to boot, Noyes spent spring 1924 searching for an expert in the field.

He found one in John J. Abel, a sixty-seven-year-old professor of pharmacology at Johns Hopkins University. An 1883 graduate of the University of Michigan, who had gained his M.D. degree from the University of Strasbourg in 1888, Abel had gone on to make a name for himself in the isolation of endocrine secretions, including adrenaline and pituitary extracts, from the suprarenal and pituitary glands, in particular. The purifying of insulin, the hormone secreted by cells in the pancreas, seemed tailored to Abel's skills. Would he be willing to spend part or all of the next academic year working on this problem? Noyes wanted to know. To entice the biochemist, Noyes dangled before him the prospect of an unlimited supply of "a high grade of insulin preparation." The bait worked. "Will attack insulin," Abel wired Noyes in July 1924, and began drawing up a list of chemicals and other supplies, ranging from white mice to rabbits and frogs, needed for the upcoming "attack" on the insulin problem.

Accompanied by his wife and his assistant, one Dr. E. M. K. Geiling, Abel spent four months at Caltech, from October 1924 to February 1925. Noyes cautioned him in advance about what to expect in the way of support. The Institute's library "is weak on the biochemical side," Noyes told him in one letter, adding that Abel ought to consider bringing "any special apparatus or books which we have not or cannot quickly get; for we have not much of any biological equipment." As so often happens in experimental work, Abel and Geiling had little to show for their efforts during their first stay in Pasadena. But two weeks before returning home, they found evidence that insulin was an unstable sulfur compound. A year later, Abel was to isolate insulin in crystalline form at Johns Hopkins. Even so, unmasking the structure of the insulin molecule remained an insoluble problem in protein chemistry well into the 1950s.

Events moved more swiftly with regard to establishing a broader curriculum at Caltech. In January 1925, Millikan applied to the Rockefeller-funded General Education Board for funds to support a wide range of Caltech activities, from physics and mathematics to organic chemistry, engineering research, and geology. Funds to add a department of biochemistry and biophysics at the Institute figured in the proposal as well. From the outset, it was clear that realizing this vision would not come cheap. In all, he requested $137,500 in support for each of the next ten years, an amount equivalent to 5 percent annual interest on just under $3 million.

GEB executives had a more modest sum in mind. They offered to underwrite one-third of the $1.35 million endowment fund Caltech needed to maintain its present activities and to upgrade organic chemistry—the first two items on Millikan's budget. Under the terms of the 1925 agreement, Caltech's trustees promised to raise $900,000 over the next three years, while the GEB pledged $450,000. The board did not act on the request to add a biochemistry and biophysics department, "leaving the question of future expansion for consideration at a later date." It took Caltech two years to raise the $900,000.

Millikan pushed for a Caltech medical school to the bitter end. The showdown came in February 1925, on the eve of Abel's departure from Pasadena. Abel, accompanied by Noyes, stopped by Millikan's house, intending to review Caltech's expansion on the biological side. Nothing of the sort happened. Instead of discussing the pros and cons of establishing a biochemistry and biophysics institute on the campus and a medical school in Los Angeles, of offering two years of premedical courses at Caltech and two years of clinical work somewhere else, Millikan, Noyes later reported, "would talk of nothing but the grand ideal of locating the whole undertaking in Pasadena."

Millikan's unbridled enthusiasm for jumping into the medical side of biology in 1925 sent chills down Noyes's spine. Picking his words with care, Noyes spelled out for Millikan himself in a memorandum how Caltech's future looked to him. He could see setting up an exchange program with a local teaching hospital, providing medical researchers with experimental animals, and establishing a comprehensive medical center in the community. But Noyes drew the line at getting into the medical business. "The community and country needs better rather than more doctors," he advised Millikan, adding, "The question of a Medical School in Southern California can best await the development of a Teaching Hospital which is essential to carry out the best modern practice in medical education." In recounting the story behind his memorandum to Hale later on, Noyes remarked drily, "I am hoping this statement may have some curative effect on him as an antitoxin serum." It seems to have done so.

Nothing was done about biology until 1927. Meanwhile, Caltech occupied itself with other matters. In 1925, it established a department of geology and paleontology, under John P. Buwalda of the University of California at Berkeley. The following year, it mounted an intensive (and successful) campaign to secure financial backing from the Daniel Guggenheim Fund for the Promotion of Aeronautics to found an aeronautical laboratory. On the financial side, its campaign to raise $900,000 in endowment funds crept along. By 1927, Caltech was still several hundred thousand dollars short of meeting the GEB's 1925 conditions. That spring, the GEB sent word that two members of the board, President Rose among them, wanted to visit the campus on March 24. The announcement from New York sent shock waves through the Pasadena campus. It also brought biology back into the picture in a big way.

Caltech ran "the risk of losing their interest," Millikan warned school officials in early March, if it did not raise its share of the endowment money by the time Rose set foot in Pasadena. Besides, Rose was slated to retire the following year—all the more reason for them to "act now," not later. As Millikan reminded his own board, the GEB had talked often of its willingness to make "another proposition." He added,

> There are at least six men in Southern California who have expressed their desire and ultimate intention of having a part of importance in the development of the Institute. If they, in view of the opportuneness of the present time, could see the way of translating now their desire and intentions into definite commitments, six bequests at $350,000 each would almost certainly bring a million in endowment funds from the General Education Board.

Among those six was Allan Balch, a newcomer to Caltech's board of trustees. Balch came to see Millikan in his office in late March and authorized him to tell the GEB that Caltech had met its 1925 grant conditions. Indeed, Balch offered to put up $400,000, if necessary.

Caltech's scientists wasted no time asking the GEB for more money. In their negotiations with it, Hale, Millikan, and Noyes had a distinct advantage over other schools: an unbeatable track record in the physical sciences. The fact that 25 percent of the country's National Research Council fellows in physics and chemistry had selected Caltech, that H. A. Lorentz, the grand old man of European physics, had declared Caltech "the world's center for research work in physics," and that Caltech led the nation's top fourteen research institutions in major physics publications pleased GEB officials.

Millikan figured he needed to raise an additional fund of $4 million—$2.6 million for endowment, $1.4 million for building costs—"to build up from the bottom" biology and to strengthen research work in the physical sciences. He wanted the GEB to match Caltech dollar for dollar, up to $2 million. "If we had up our sleeve the commitments which would enable us to cash in at once so far as their [GEB's] end of it was concerned," he told the members of his executive council in 1927, "this center here in Southern California would begin to move forward this spring in a way which would be inspiring, not only to the southland, but to the whole of the country." What Caltech's leaders had up their sleeve was Columbia University's distinguished biologist Thomas Hunt Morgan.

Morgan received his Ph.D. in embryology in 1890 from Johns Hopkins University. A naturalist at heart, Morgan had then worked on a wide range of plants and animals, large and small, including coleus, mariposa lilies, frogs, homing pigeons, earthworms, aphids, and crabs; all forms of marine life interested him. But the studies that he went on to and that brought fame to him and his Columbia research group concerned heredity. Starting with the study of chromosomes in aphids, he graduated to genetic experiments using chickens, rats, and mice, before turning finally to *Drosophila* (more commonly known as the fruit fly) for laboratory work. It was a question of mutations, and in an effort to induce them in animals, Morgan subjected the flies to temperature increases and X rays, but the results were inconclusive. Then, in 1910, he noticed a male fly with a new eye color—white, instead of red—in one of his culture bottles. Historians generally agree that his first paper on the white-eyed *Drosophila* mutant, which appeared that summer in the magazine *Science*, opened the door to modern genetics.

Caltech officials first sounded Morgan out in winter 1927 about the possibility of organizing work in biology in Pasadena. Morgan took an instant shine to the idea. His name came up again in conversation with GEB officials in Pasadena in late March, when the talk turned to bringing in "an outstanding man in biology." Several weeks later, on April 6, Hale and Millikan called on the GEB's President, Rose, in New York. Hale urged Rose to take a look at a recent article by Morgan entitled "The Relation of Biology to Physics," in *Science*. Not only did Rose read it; he went uptown a day or so later and had a long talk with Morgan. While they were at it, Morgan showed Rose around Columbia's physics laboratory. By mid-April, talks between Morgan and Caltech's scientific trio had shifted into high gear.

Before the month was out, Morgan announced he was ready to cut his ties with Columbia ("He said *definitely* that he will come to us," Noyes informed Hale), provided Caltech would put up a biology building and give him an $80,000 annual budget. "I can't tell you how much I should enjoy having you as one of our little partnership," Hale wrote to Morgan upon hearing the news. "There is a special place reserved for you before the fire in my library, where we periodically talk over our plans." "The participation of a group of scientific men united in a common venture for the advancement of research fires my imagination to the kindling point," Morgan wrote back, returning the compliment. Caltech would get the funds, Hale assured the biologist, "by any hook or crook."

Jubilant, Caltech officials reported back to the GEB that they had found the Columbia geneticist "greatly interested in our scheme to aid in the development of a rational biology based on physics and chemistry." If the GEB can "make the gift we requested we will put the thing through," Hale promised Rose.

While Morgan "has not finally agreed to accept," Hale added, he "thinks the California Institute the only place to develop the scheme, partly because of the physics and chemistry and the small number of students, and also because there is no existing department of biology and no faculty members or traditions stand in the way of an entirely new scheme." To say that Morgan hadn't yet agreed to come was a nice move; it put the ball right back in the GEB's court.

Up to this point, Rockefeller Foundation officials had been insisting that "the needs of the existing depts be attended to *first* and that biology be then begun," and they urged Millikan to organize his budget request along these lines. In keeping with their philosophy, in April 1927 the GEB informally offered to contribute $55,000 annually for three years for programs already under way, $25,000 for biology (Millikan had requested $75,000), together with a matching grant of $800,000 for endowment. The appropriation in biology, they explained, was "to make the necessary arrangements to secure the key man and initiate the plans for development." Later on, when plans matured, the board would consider additional appropriations. But the lure of Morgan's name accelerated the maturation process.

In mid-June, Caltech's board of trustees met, put its stamp of approval on Morgan's conditions, and formally invited him to join the faculty. Afterward, Millikan sent Morgan a copy of the board's resolution, along with a brief note stating that he was leaving for the East Coast at the end of the month and would like to see him in New York the following week. At their meeting on July 4, Morgan recited the conditions of his appointment, described the research fields he wanted to see developed at Caltech, and spoke of the need to find "suitable men." By the time Millikan rose to leave, Morgan later recalled, "an understanding was reached."

The following day Millikan dispatched a letter to Allan Balch, the trustee whose $50,000 contribution at the last minute had helped Caltech meet the GEB's 1925 grant. Born in upstate New

York and trained in electrical engineering at Cornell University, Balch headed for the Pacific Northwest following graduation. In 1896, he moved from Portland to Los Angeles, where he learned the ropes of managing a power company. A pioneer in the development of electric power in southern California, Balch teamed up with William G. Kerckhoff, Henry W. O'Melveny, Kaspare Cohn, and Abe Haas to establish three imposing utility companies: the San Gabriel Light and Power Company, which tapped the waters of the San Gabriel River to supply Los Angeles with light and power; the San Joaquin Light and Power Company; and the Southern California Gas Company. After the sale of these utilities in 1927, Balch and his wife, Janet Jacks Balch, poured their time, energy, and money into developing the arts and education in southern California, including Caltech.

At the time of Caltech's negotiations with Morgan, Balch was aboard ship en route to New York, scheduled to arrive there July 15. In his letter to Balch, Millikan touted Morgan as "without doubt the foremost biologist in the United States, perhaps in the world," who had just been elected president of the National Academy of Sciences by his colleagues and to whose "extraordinary work on chromosomes" Balch must often have seen references. Morgan had "definitely and irrevocably accepted" their offer, Millikan told Balch, and it remained only to match the GEB's $1,050,000. "I think you will agree that something momentous has happened as a result of our interview of last March," Millikan added, before launching into the real purpose of the letter. Would Balch be interested in providing the building? It took the industrialist less than a week to make up his mind.

After the boat docked, Balch met with Arthur Fleming, the president of Caltech's board of trustees, and assured him that he personally would take care of Morgan's needs. Fleming understood Balch to mean $2 million and telegraphed Hale to this effect. Although skeptical, Hale relayed this information to Millikan as well as to Morgan. Nobody had the nerve to ask Balch directly what exactly he had promised Fleming. Believing it was

$2 million, Morgan on August 1, 1927, formally accepted Caltech's offer to develop a department of biology in Pasadena. Architects were called in, and plans made to construct two buildings—one for genetics and the other for physiology, developmental biology, biophysics, and biochemistry. The sad truth—that Balch had pledged $1 million, "but nothing more"—dawned on Caltech officials only in October, and by that time it was too late for Morgan or the Institute to back out of their deal. Balch, it turned out, had simply told Fleming he would try to get a friend of his to help pay for the buildings.

Morgan exploded. "It seems incredible from a business point of view," he rebuked Hale, upon learning how matters really stood, "that we could have gotten into such a muddle—that commitments should have been allowed to go so far without a specific statement of guarantees. There remains nothing to do now but put our shoulders to the wheel and try to get out of the rut." Morgan showed the way by getting the trustees to guarantee that the income on the $2 million in hand "be set aside for salaries" in his department. "Even if we have to sit on the campus or wander through the Groves of Huntington [the Henry Huntington Library and Art Gallery]," insisted Morgan, staffing the department "has first importance because it is going to affect the kind of men we can hope to approach at the start which will determine the entire future of the department." In a sense, Morgan was not asking for any special favors: he was simply pressing Caltech school officials to live up to their promises. As it turned out, Morgan's immediate needs were purchased at the expense of organic chemistry and other things on Millikan's own wish list.

Caltech quickly scaled back its construction schedule. "It may be a long time before we can erect your whole group of laboratories," Hale warned Morgan. Because it still had to provide for Morgan's coming, the institution reluctantly put up the money itself to begin construction on the building. Morgan himself spent hours poring over laboratory plans, dividing up the floor space, designing basement windows, approving the preliminary layout of outlets and wiring, and selecting cabinets, sinks, tables, and other laboratory equipment. By Thanksgiving 1927, Caltech was ready to break ground for the first unit of the new biology building, located at the west end of the campus.

Behind the scenes, Caltech officials scrambled frantically to reach a new accommodation with the GEB. When Millikan returned to Pasadena from a League of Nations meeting in Geneva in October 1927, he requested a second million from them on the same terms as before. He received a courteous, but correct, reply. The board would accept a new application when Balch had actually turned over his promised gift, not before. As to the proposed terms, GEB officials had bent their own rules in matching Caltech dollar for dollar; the board now expected Caltech to contribute two dollars for every new Rockefeller dollar pledged, meaning Caltech needed to raise two million the second time around, not one. That November, Balch himself visited the GEB office in New York and listened attentively while a staffer explained to him "the interest of the Board in aiding institutions to add to their clientele of donors."

Balch found his donor in William George Kerckhoff. A resident of Los Angeles for forty-nine years who lived next door to Norman Bridge and was business partners with Balch, William Kerckhoff, seventy-two, offered early in 1928 to pay $1 million toward the cost of putting up Morgan's buildings, provided they were named after him. Millikan told Morgan, "I have spoken to Mr. Kerckhoff and find that he is not particular about the form if the laboratories are named after William G. Kerckhoff." Academic buildings are traditionally named after the men and women who give them, and Kerckhoff wanted that recognition.

Balch did not. Six months after pledging $1 million in endowment for biology, on February 3, 1928, Allan Balch formally turned over to Caltech $1,050,000 worth of Southern California Gas Company bonds. He gave the gift anonymously, telling Millikan that he did "not wish to have his name attached to the department or the buildings." Nor did the GEB impose any such stipulations on its own gifts. Since these two donors (Balch, the Rockefeller Foundation) did not insist on having the biology laboratories named after themselves, Millikan was able to assure Kerckhoff the building would bear his name. Of the philanthropists who have contributed to Caltech's development, Balch is, ironically, among the least remembered, because his name is not attached to any building.

Kerckhoff's million counted for half of the $2 million condition imposed by the GEB on Caltech. It took the institution less than a year to raise the remaining million. In all, Hale, Millikan, and Noyes raised $5 million, principally for biology, in a span of something less than two years, and captured one of the world's leading biologists in the bargain. In summer 1928, Thomas Hunt Morgan arrived in Pasadena to take up his new duties as head of the biological laboratories.

SEVEN

◆

Earthquakes

In the course of historical or statistical study of earthquakes in any given region it is frequently desirable to have a scale for rating these shocks in terms of their original energy, independently of the effects which may be produced at any particular point of observation. On the suggestion of Mr. H. O. Wood, it is here proposed to refer to such a scale as a 'magnitude' scale.
—CHARLES RICHTER, 1935

THE SAN ANDREAS FAULT, meeting place of two of the great, ever shifting tectonic plates that make up the planet's surfaces, runs along the California coast from the Mexican border to San Francisco Bay. But between the two world wars, the existence of tectonic plates was still a far-off discovery of global geology, a subject then popular mostly in Europe. Local earthquakes, on the other hand, were a subject of immediate interest to anyone living in the Golden State, and that interest led to seismology at Caltech, to a marriage in the thirties between two different traditions—a European interest in global earthquakes and an American concern for the stability of the earth beneath us.

Between the great San Francisco earthquake in 1906 and Caltech's modest beginnings in earthquake study in Pasadena in 1921, many had called for the development of a program in seismological research and few had heeded the call. California researchers found the business community hostile to studies that might stifle growth. Oil geologists were reluctant to share technical information with academic geologists, and state university officials in Berkeley were unsympathetic to a geophysical institute where seismological investigations could be carried out in the southern part of the state. In the end, it took more movers and shakers than just the San Andreas Fault to bring seismology to southern California.

When seismology ultimately did come to Pasadena, it centered on three men—Beno Gutenberg, Harry Wood, and Charles Richter. German-born and Göttingen-trained, Beno Gutenberg brought to Caltech the European tradition of viewing seismology as a research tool. Rigorously educated in physics and mathematics, Gutenberg used earthquake records to investigate the physical properties and structure of the earth's interior. Earthquake instruments installed in seismological stations around the world provided data for his analysis. The globe was his scientific laboratory. For Gutenberg, an earthquake anywhere on the planet rang the earth like a bell, emitting seismic waves. A study of these waves and their progress from epicenter to detecting station would draw a picture of the composition of the globe.

Gutenberg had made his mark in seismology in 1912, long before Caltech was even a twinkle in Robert Millikan's eye. By the mid-1920s, Gutenberg had published more than a score of important scientific papers, edited and written several chapters for the monumental *Handbook of Geophysics*, and produced two textbooks—an introduction to seismology and a volume on geophysics. By 1930, Gutenberg ranked as high among seismologists as among geophysicists. But his ultimate dream, a professorship at Göttingen, was beyond his reach despite all his distinguished accomplishments. Beno Gutenberg was a Jew.

Harry Oscar Wood, who started out studying the properties of minerals, took a much more pragmatic view of the world than did Gutenberg. In this he resembled his American colleagues, most of whom worked on California earthquake records in the hope of finding a solution to the "California problem," as they called regional earthquakes. Members of the American seismological community, formed largely of geologists, mining engineers, and astronomers, all saw their research rather narrowly. To a seismologist like Wood, Gutenberg's global problem shrank to the size of southern California.

Indeed, Gutenberg and Wood made an unlikely scientific pair: Wood, the outsider in the academic world, a veteran of the San Francisco quake and the expert on the seismic history of California; and Gutenberg, the erudite university professor of the straitlaced German school, sufficiently unorthodox to believe in continental drift, and internationally respected for his ability to decipher instrumental recordings of earthquakes. Nevertheless, Wood and Gutenberg worked side by side in Pasadena's seismological laboratory, trading seismograms and technical information daily. In bringing these two men together in Pasadena in 1930, Robert Millikan set in motion a chain of events that led to Charles Richter's development of his famous earthquake magnitude scale in the early 1930s.

Charles Richter was the youngest of the trio. He took his Ph.D. in theoretical physics under Paul Epstein at Caltech, became Wood's assistant in 1927, three years before Gutenberg's arrival, and then spent the next several years measuring and filing seismograms. Much of the work was tedious and mechanical, and Richter's contacts with seismologists, aside from Wood, remained limited. A theoretical physicist who fell into seismology more or less accidentally, Richter desperately needed a scientific mentor. Gutenberg fit the bill; indeed, if Gutenberg had not come to

Richter, Richter would have gone to Gutenberg. As things turned out, they spent thirty years under the same academic roof.

Like much of the history of seismology in America, the Wood-Gutenberg-Richter story has its roots in the 1906 San Francisco earthquake. Harry O. Wood hailed from Maine. He had come west to Berkeley in 1904 after several years of undergraduate and graduate work at Harvard in mineralogy and geology, finding work as an instructor in the geology department at Berkeley, thanks to its energetic leader, Andrew C. Lawson. The 1906 earthquake tossed Wood, like many another, out of his bed and into seismology. Berkeley's Lawson long played a role in guiding Wood. Director of the State Earthquake Investigation Commission, set up to study the 1906 tremor, Lawson asked Wood to examine all of its detail and damage within the city. Wood went into the exercise a field geologist and came out a seismologist. Like Lawson, he lobbied for a seismographic station on the campus, and when Berkeley's seismographic station started operations in 1910, under the direction of the geology department, Wood joined the basement-room operation. He did everything from analyzing the seismograms to publishing the new *Bulletin of Seismographic Stations*.

Even so early in the century, credentials mattered. Hampered by the lack of a Ph.D., Wood left Berkeley in 1912 for a position as research associate at the Hawaiian Volcano Observatory. In addition to caring for the observatory's seismic instruments and analyzing their data, he published papers on the eruptions of Mauna Loa of 1914 and 1916, among other aspects of vulcanology. But Hawaii did not suit him. He felt out of touch. Moreover, he wanted to run something, and there was no chance of that at the observatory. His five-year stint with the observatory, Wood later said, left him only with "a deep coat of rust, scientifically."

While in Hawaii, Wood laid the groundwork for his return to California. In two 1916 papers, he stressed the importance of taking a regional approach to the study of local earthquakes. The first one collated and cataloged all the known California earthquakes. In his second paper, "The Earthquake Problem in the

Western United States," Wood spelled out in detail a research plan that called for the establishment of a network of seismic stations up and down California and in neighboring states. Significantly, he suggested that the plan be tested on a modest scale in southern California. These papers later served as blueprints for the seismology program at Caltech.

Wood singled out southern California for two central and sufficient reasons: the region possessed no recording instruments, and he expected the next large earthquake to occur there. The 1857 earthquake along the San Andreas Fault had been southern California's last great shock. The next quake, if properly instrumented, was sure to pay off with a mother lode of invaluable data. He recognized, of course, that putting the plan into practice would contribute to the science of geophysics, but this was not his principal motivation. Wood was not interested in recording distant earthquakes. To use southern California as a laboratory for "the instrumental study of strong earthquakes," when Los Angeles and the surrounding cities lacked the means to monitor even local ground motion, seemed senseless to him. The seismology program was not a way to learn more about the physical condition of the earth's interior. It was a way, perhaps, to save lives.

Wood's research program stressed the need for a new generation of instruments; there could be no hope of measuring short-period, local earthquakes with instruments devised to measure long-period, distant earthquakes. Drawing on his background as a field geologist, Wood also pointed out the need to make extensive field measurements, including triangulation studies that would measure evidence of shifts in the earth's surface, to verify known faults, and to identify new ones. The fieldwork required to locate weak shocks figured prominently in Wood's plans. Like most of his contemporaries, Wood was convinced that if geologists could identify the active faults associated with these weak shocks, they could "deduce . . . the places where strong shocks are to originate, considerably in advance of their advent." They supposed that big shocks follow weak shocks.

The compelling reason for setting up branch seismological

stations was therefore to detect and register weak shocks systematically. The stations would have spring-loaded masses, whose vibrations in response to seismic tremors would be recorded by a needle scratching a mark on a slowly rotating drum covered with smoked paper. But the routine registration of hundreds and hundreds of small California earthquakes did not confirm Wood's hypothesis that weak shocks "should be telling harbingers of strong ones." Instead, the tremors seemed to obey no recognizable or predictable pattern. Nevertheless, the lure of that elusive idea—it seems so logical—drives seismological research to this day.

During World War I, Wood returned to the mainland from Hawaii and joined the army, with the rank of captain in the Engineer Reserve Corps. He was assigned to the Bureau of Standards in Washington, where he worked on the design of a recorder capable of detecting and registering distant cannon fire. Save for the much shorter periods, the vibrations set up in the earth by artillery resemble those of local earthquakes. Making use of the fact that certain crystals, called piezoelectrics, develop an electric voltage when squeezed, Wood devised a piezoelectric seismometer. At the close of the war, he remained in Washington, working for the National Research Council and talking up his plan to study earthquakes in southern California. Knowing that the NRC's founder, George Ellery Hale, also headed an astronomical observatory near Los Angeles, Wood had first told him in 1917 about his idea, and again in 1919. Keenly aware of the local seismic risk, Hale now suggested that Wood talk with Arthur Day, head of the Carnegie Institution's geophysical laboratory in Washington, D.C. Day turned out to be already familiar with Wood's plan. He passed it on to John Merriam, a Berkeley paleontologist and geologist, who with Hale's support had left university life in 1920 to become president of the Carnegie Institution of Washington. Merriam gave Harry Wood the chance he craved.

You do not have to live in California to be interested in earthquakes, but it helps. Not only did Merriam qualify on that count—he knew California well—he also knew that the Carnegie Institution had played banker to Lawson's 1906 earthquake investigation at Berkeley, and he knew that Berkeley's geologists had no territorial ambitions south of the Tehachapi Mountains, which separate northern California from southern California. Thus it was possible to start up a new venture in the south.

At that time, Merriam trusted Hale. Not only that; Merriam—and Carnegie—had long supported Hale's Mount Wilson Observatory. If Hale thought Wood's proposal had merit, Merriam could do worse in his new job than encourage it. Also, Merriam had more than a casual interest in Hale's newly christened godchild, the California Institute of Technology, located several miles due south of the observatory's Santa Barbara Street headquarters. He had reserved $150,000 for physics and chemistry research at Caltech, and when geology's turn came, he even had someone in mind for the chairman's job. In the spring of 1921, Merriam formed an advisory committee on seismology, under Arthur Day's direction. The committee, comprising five geologists and two physicists—Millikan, then on the verge of becoming head of Caltech, and John C. Anderson, a Mount Wilson astronomer—was to oversee the Carnegie's new West Coast program. In one move, Merriam satisfied Hale's concern that something be done to monitor local earthquakes, repaid Hale for his part in making him president of the Carnegie Institution, and set the stage for the beginnings of geology at Caltech.

Southern California's first seismological program began operation in Pasadena in June 1921 under Harry Wood's direction. For the next six years, Wood ran the project from an office at the Mount Wilson Observatory.

Wood did not like publicity, at least in the beginning. This brought him into conflict with Ralph Arnold, a local geologist and a member of Merriam's advisory committee, who wanted to involve the public as much as possible in Wood's earthquake research program. Among other things, he proposed that citizens keep a log of local shocks. Day seemed resigned to Arnold's role.

"I suppose in general that publicity is the daily food of that part of the country, and no undertaking may hope to thrive without it," Day wrote Wood, who had complained about Arnold's "boosterism." All the same, Day promised to talk to Merriam about Arnold's publicity work, adding, "He knows California better than I and may have some suggestions." But Merriam had put the knowledgeable and well-connected geologist on the committee precisely because he needed Arnold's talents to wage a grass-roots campaign to make the Carnegie's earthquake research project respectable. Before the year was out, Wood had given a series of talks on the earthquake problem in southern California— all planned by Arnold—to scientific and social groups, including the Twilight Club, an exclusive private club to which Hale and other prominent Pasadena men belonged.

In his talk before the Twilighters, Wood attacked the prevailing attitude of Californians toward the study of earthquakes, saving his harshest criticism for those "men who fear that greater knowledge of these phenomena will prove that region dangerous for human habitation and so injure them through its tendency to deter immigration and slow down the commercial development of the region." Where had the occurrence of earthquakes triggered a mass departure of the population? "A less brazen people," he told the group, "would hide in shame—but they—not they: 'we don't have earthquakes—just a little shake now and then—just keeps us alive you know.' " For a reluctant publicist, Wood played his audience well. First he shamed his listeners; then he appealed to their regional pride. "Look at Mexico," he said, "with one station better equipped than *any other* in North America." Wood revealed his plans to catch up with the Mexicans:

We may say broadly, without calling anybody names, that in the United States there are few well equipped seismometric stations, none whatever of the first class—and many of those existing are decidedly of third class grade or lower. . . . We need stations more intelligently located, better planned, better equipped, and thoroughly well cared for by men . . . whose chief interests are seismology and whose training in theoretical and practical physics and geology has been adequate and conducted with the practice of seismology in mind. These men must come from the rising generation.

Wood followed his own prescription for success closely. He recruited Richter, a theoretical physicist, and Hugo Benioff, an astronomer turned seismologist, for the project; and he raised no objections to Gutenberg's appointment at Caltech. As for a first-rate central station, Wood got that too, as we shall see.

In the meantime, Arnold more than met Merriam's expectations. His campaign to make earthquake research acceptable ranged widely. He organized a local geology club; he put Wood in touch with the region's oil geologists. He persuaded the town elders in Riverside and elsewhere to pay for the monitoring stations Wood had planned for their communities. With Arnold's help, Wood and the Stanford geology professor Bailey Willis, also a member of the advisory committee, compiled and published in 1923 a detailed map of fault lines up and down California. Never before had the system of fault lines in the Los Angeles vicinity, including the Garlock, the San Jacinto, and the Imperial faults, and the location and activity of the great San Andreas Fault, been spelled out with such authority and precision.

Even more than publicity, the project needed the right instrument for recording nearby earthquakes. Fortune favored Wood in the person of John Anderson, one of Hale's ablest astronomers. As part of Throop College's defense effort during World War I, Anderson had worked on submarine instruments sensitive enough to detect and record short vibrations. Anderson's war-honed skills matched Wood's peacetime needs. The Wood-Anderson collaboration began immediately after Wood had settled into his office in 1921.

Wood wanted an instrument responsive enough to record shocks having a period (the back-and-forth time of one oscillation) varying from 0.5 to 2.0 seconds. Seismometers for recording distant earthquakes are typically designed to respond to longer-period oscillations, because much of the sharp, rapid shaking of nearby earthquakes dies away at long distances. In the case of Berkeley's station, the Bosch-Omari and Weichert instruments had periods of 15 and 6 seconds, respectively. In the early twenties, because of these longer periods, instruments on the Atlantic seaboard could measure the time and place of California shocks better than comparable instruments located in California.

By the fall of 1922, after several false starts, Anderson and Wood had designed a reliable, compact, portable instrument that, when placed vertically, consistently recorded the east–west and north–south components of the earth's motion during an earthquake. Wood described his partner's part to Day: "Anderson has contrived a very light, very strong and rugged, quick-period mechanical vibrator, controlled wholly by the torsion of a very small tungsten wire, damped magnetically, which . . . will register . . . all three components of motion." Wood was overly enthusiastic about the instrument's capability. In practice, the Wood-Anderson torsion seismometer was an ideal instrument for recording the earth's horizontal movements over a short distance during an earthquake. However, it didn't do well recording the ups and downs of an earthquake.

Shortly after Gutenberg arrived from Germany, in 1930, Hugo Benioff, Wood's assistant, designed and built a vertical seismometer to meet Wood's needs. Routine recording of local shocks by means of Benioff's instrument began in 1931. Wood, typically enthusiastic, was predicting that the new vertical-component seismometer would surpass any existing vertical instrument then in use for the recording of distant earthquakes as well. And it did. Both the Wood-Anderson and the Benioff instruments have since become standard equipment in seismic stations around the world.

The first Wood-Anderson instrumental records were made in December 1922; the first extant records date from mid-January 1923. "There ain't no other seismograph worth talking about than ours," Anderson wrote Wood after showing off their invention to Merriam, Day, and others at Carnegie's Washington office a year later. Indeed, the Wood-Anderson torsion seismometer did more than its creators had intended. Wood had wanted a short-period instrument to register local earthquakes. But when the instrument was put to the test in 1923, he discovered that it also registered the first phases of distant earthquakes. Wood had unwittingly altered the course of his own program.

This development should have delighted Caltech's chief executive officer, who had little interest in a research program aimed at a purely local problem. The surviving records make clear that Millikan hoped to broaden the scientific mandate under which Day's advisory committee operated. But Day would not budge. "The [Carnegie] Institution is committed to the study of the local disturbances of California for the next few years, . . . rather than to the teleseismic [distant earthquake] problem," he told Millikan, who had proposed that more attention be paid to the latter. Skeptical of Millikan's professed interest in the committee's project, Day asked Wood how much help they could count on from Caltech. "Dr. Millikan has too many irons in the fire . . . to hold much interest" in the local work, he replied. "As for the world problem," Wood continued, Millikan *"might* take this up as part of the work of the Institute. . . . He might co-operate on the teleseismic side, or he might restrict his co-operation to theoretical questions."

By the spring of 1924, the experimental torsion seismometers installed in the basement of the Mount Wilson Observatory office and the physics building on campus had recorded dozens of earthquakes, near and far, including the initial short-period phases of the devastating Japanese earthquake of September 1, 1923. That Wood had recorded this event on an instrument designed to register local earthquakes was particularly exciting. Gutenberg, then still at Göttingen, held up the publication of his book on the fundamentals of seismology long enough to insert a diagram of the apparatus. After studying in detail a seismogram of the South Pacific earthquake of June 25, 1924, sent by Wood, Gutenberg, the expert in worldwide earthquakes, wrote back, "The seismogram is indeed extraordinarily interesting. . . . I look forward to further communications from you on the success of your instrument." Wood did not fail to communicate.

Closer to home, the Berkeley seismologists, led by a Jesuit priest named James B. Macelwane, urged school officials to buy the new Wood-Anderson instrument for the campus station. When Macelwane learned in 1925 of his appointment as professor of geophysics at Saint Louis University, in Missouri, he ordered several torsion seismographs for his new laboratory. Five years later, the Berkeley station also purchased a set of the new horizontal-component instruments for the purpose of registering local shocks. In 1930, thirteen cities in the United States, and one overseas, had Wood-Anderson seismographs.

After so brilliantly recorded local and distant shocks, Caltech made its first serious bid for a role in Wood's scientific work. The construction of a permanent, central seismological station in Pasadena set the stage. Day wanted to build it and three other stations, each equipped with a set of short-period torsion instruments, as soon as possible. The sites initially picked—Riverside to the east, and La Jolla to the south—would form with Pasadena a convenient triangle for calculating the location of local shocks. It was not easy to put the apex of this system into place. Although Carnegie's Merriam and Day wanted to find a permanent laboratory for Wood, they had no intention of paying for it.

Day tried to turn the problem over to Hale. Now the Carnegie Corporation, acting on the advice of Merriam, had voted to give Caltech $25,000 to start a geology program. Hale returned the problem to Day with the proposal that Caltech would guarantee the site and building if the Carnegie Institution would "permit the [seismology] work to be conducted in the future as an integral part of the Department of Geology and Geophysics." He pointed out to Merriam, who was still active as a geologist, that donors "would be especially attracted by the possibility of developing seismological research as a vital factor of the Department's study of the geology of Southern California." The Caltech geology department existed in name only. There were no geophysicists, let alone donors, in sight. None of this deterred Hale from establishing Caltech's claim to Wood's territory. If control of the program passed from Santa Barbara Street (where the Carnegie organization had its Pasadena offices) to Caltech, the unlettered Wood would no doubt be eased out.

Wood spoke favorably in public about Hale's plan, but to Day, in private, he vented his feelings. "We must keep the merger on a strictly *cooperative* basis and not lose our identity nor surrender control of our funds," Wood stressed in one of his weekly letters to Washington. "If we had only Dr. Hale to deal with this warning would not be in the least necessary," he added, "but a tendency to absorb and appropriate us into the California Institute . . . must be watched for and prevented." By insinuating that Millikan put Caltech's interests first, Wood was hoping to extract from Day a promise that the local earthquake program would remain under his—Wood's—direction. Millikan, for his part, played only a minor role in Hale's negotiations with the Carnegie Institution over the control of seismological research in the southland. As the country's newest Nobel laureate, Millikan was far too busy giving lectures in the Midwest and elsewhere, and raising money for a national research fund, to do much more than lunch with Merriam and Day during National Academy meetings in Washington. He left the future of Caltech's graduate program in geology to Hale and A. A. Noyes, the head of Caltech's chemistry program.

In 1925, John P. Buwalda accepted Caltech's invitation to head the school's graduate research program in geology. This was no coincidence at all, since Buwalda had been Merriam's student at Berkeley. The seeds of cooperation that Hale sowed in 1924, nourished by the building pledge the Caltech trustees Arthur Fleming and Henry Robinson signed in 1925, bore fruit the following year when Millikan formally invited the Carnegie Institution to conduct its earthquake research in the Institute's new seismological laboratory, located in the foothills of the San Rafael Mountains, a short drive from the campus. Under the terms of the agreement, Millikan gave the Carnegie Institution exclusive rights to conduct research in the building; Merriam, in turn, invited Caltech to cooperate in the research, provided that Day and his counterpart at Caltech, Buwalda, gave their consent. Most important, Carnegie supplied the funds to run the operation. The document Buwalda and Day finally initialed in Pasadena on June 10, 1926, purchased time for Millikan and his faculty to judge the worth of Wood's research.

Early in January 1927, Wood left his temporary quarters at the observatory office and moved into the new seismology laboratory. The time had come to go earthquake hunting in earnest. By 1929, six outlying stations, all within a 500-kilometer radius of the central station in Pasadena, were in place and working. Each boasted a pair of horizontal-component torsion instruments, recording drums, and radio-timing equipment. Records flowed weekly to Pasadena for photographic processing, registration, and interpretation.

Technology now led the science. Scientists at the Institute, pointing to the mounting number of near- and distant-earthquake records (the number of local shocks in 1928 alone exceeded 150), demanded more research on the theoretical problems of earthquake waves. Wood paid no heed. "The most serious criticism I ever heard anyone make of your work in Pasadena," Day pointedly told Wood in 1929, "was substantially this—that you have the best instruments in the world but are doing nothing with them."

Buwalda headed the list of local critics. Ambitious for his young department, restless with what he saw as the parochial outlook of Carnegie's program, he pressed Millikan to seek joint administration of the seismology laboratory. By 1929, the Caltech-Carnegie agreement of 1926, which gave the school no say in the running of the laboratory, looked very unattractive to him.

Besides quarreling over rights, Buwalda revived a much older argument, having to do with the laboratory's name. From the beginning, Wood had insisted that the Carnegie Institution's name figure prominently on the new laboratory's stationery. Wood's design mentioned Caltech's name only in the laboratory's address. Millikan rejected the design, declared that it ought to read "Seismological Laboratory of the California Institute of Technology," and asked Noyes to fashion a letterhead. Wood objected to the result on the grounds that bills would be sent to Caltech rather than to the Institution if Caltech's name appeared under the heading "Seismological Laboratory." In January 1926, he told Day, "I think it should some way be made clear that the Carnegie Institution is responsible for bills . . . and that I am locally in charge of its seismological activities." He wished that the school's name could be left off the letterhead altogether.

Millikan then proposed that the names of the two institutions be printed "as minor side-heads. . . . This would give coordinate recognition to both institutions, without undue prominence of either." But the payer called the tune, and in the end Millikan accepted a letterhead that made no mention of his school, save in the address. In practice, as Buwalda pointed out to Millikan in 1929, other seismological stations knew Caltech's station simply as the "Pasadena Seismological Laboratory," naming the place after the city, as seismologists habitually do.

The fight over the letterhead was a skirmish in Wood's defense of his position in the program. "I think it is in the back of Dr. Millikan's head," he wrote to Day when the matter arose, "that the Seismological Laboratory is subordinate to the newly created Department of Geology . . . and I hope most emphatically that no step will be taken which can be construed as subordinating our work to the Institute." Day also recognized that Millikan now looked upon the earthquake work as an extension of the Institute's. His new attitude stemmed from the fact that Caltech funds had been expended in the construction of the laboratory. The quarrel over the laboratory's name was the first step toward the

Carnegie Institution's exit from the program in the late thirties. Eventually, geography called the tune.

In October 1929, the Carnegie Institution organized a conference of American and European seismologists in Pasadena, including Gutenberg, in connection with a special meeting of its advisory committee on seismology. They came for scientific talks, field trips, informal discussion, and evaluation of the program. Gutenberg, who had visited with Chicago relatives earlier in order to "brush up his English," discussed the formulas and methods used in determining the speed and paths of earthquake waves as they travel through the earth. His remarks struck a responsive chord in Wood. The design of his regional earthquake program, now "on the carpet"—according to Day—rested on determining the epicenters and depths of local shocks, and neither Wood nor his assistants could match Gutenberg's experience in this area. "We need Gutenberg more than Europe does," Wood told Day after the meeting.

At the conference, Charles Richter talked at length about his work. Richter had joined Wood's enterprise in the fall of 1927, hoping that something permanent would turn up at Caltech. His love affair with Caltech had its roots in the early 1920s when he came to the campus to hear Millikan lecture on developments in modern physics. A Stanford graduate, Richter quit his job as a warehouseman and enrolled in Caltech as a graduate student in physics. He took advantage of the remarkable opportunities of the place, which then included guest lectures by Erwin Schrödinger, Max Born, and James Franck, as well as the English professor Clinton Judy's famous graduate seminar in literature, which attracted a remarkable cross-section of the Caltech community. "Not particularly social," by his own admission, Richter regularly attended these biweekly evening discussions at Judy's home, trading talks on topics ranging from the influence of scientific ideas on literature to the philosophy of John Dewey, with arguments about the practical applications of science, Nietzsche's ideas, and genetics with the likes of the physicist J. Robert Oppenheimer, the astronomer Fritz Zwicky, and the biologist Calvin Bridges. "These were almost too interesting times," he remarked many years later. Richter intended to stick around Caltech.

Millikan recommended the wide-ranging graduate student to Merriam in the spring of 1927, predicting that Richter would leave his mark on science, regardless of the field he picked. Wood hired Richter that fall to work part-time reading and interpreting seismograms in the new laboratory. When Richter finished his dissertation, in 1928, he told Wood he wanted to stay on. In correspondence with Day about raising money to put Richter on the lab's permanent payroll, it was agreed that Richter would not be "an ordinary routine worker." Day wanted to send Richter to Germany to study with Gutenberg; even better, he wanted Millikan to bring Gutenberg to Caltech. Either way, he told Wood, "I would prefer . . . to provide the aid above Richter rather than below him."

Many years later, Richter recalled that everyone at the Pasadena conference of October 1929 understood that Millikan had his eye on Gutenberg or the English geophysicist Harold Jeffreys. But November passed and nothing happened until December, when Millikan learned that Harvard planned to offer Gutenberg a job. On thirty minutes' notice, Millikan summoned Wood, Anderson, and Buwalda to his office. Two hours later, Millikan cabled Gutenberg asking if he would consider a position at Caltech.

Beno Gutenberg discovered seismology just as it was beginning to stand on its own as a branch of geophysics. Born in 1889 in Darmstadt, where his grandfather owned a soap factory, he excelled in mathematics and physics as a teenager, taught himself the rudiments of meteorology as a hobby, and planned to become a high school teacher, until one of his professors at the local polytechnic school suggested that his future lay in Göttingen. The University of Göttingen turned the amateur meterologist into a dedicated scientist. He took several courses in geophysics and many in mathematics, including ones in algebra, non-Euclidean geometry, the theory of functions, and mathematical logic, taught by Hermann Minkowski, Felix Klein, Hermann Weyl, and David Hilbert, among others. But the young mathematician put geophysics first. At the end of three years, and "after a Laboratory course and a lecture course in Seismology," Gutenberg recalled many years later, "Professor [Emil] Wiechert told me . . . I had progressed about to the limit of known results in Seismology and suggested to start with a thesis research." In May of 1911, Gutenberg successfully defended his dissertation, dealing with microseisms—he traced the source of the small continuous disturbances registered on Göttingen's seismographs to the action of the surf along the coast of Norway—turned in to the university the required 200 reprints of the published thesis, and received his Ph.D. He was not yet twenty-two.

While still working on his thesis, Gutenberg received from Wiechert, the director of the Göttingen Institute for Geophysics, an incomplete manuscript by Karl Zöppritz, a promising seismologist—and Wiechert's former assistant—who had died suddenly in 1908. Gutenberg read the manuscript with care and began to make some calculations. Zöppritz had been studying the amplitudes of elastic waves in earthquakes as a function of their depth. He had posed but not solved the problem. "I considered this paper fundamental," Gutenberg said once, "and upon my report to Professor Wiechert he suggested that the first assistant of the [Geophysical] Institute, L. Geiger, and I finish and expand the manuscript for publication." They did so. Gutenberg worked out the formulas for calculating amplitudes by means of Zöppritz's method and afterward, with Ludwig Geiger's aid, used the method on real earthquake waves.

Gutenberg then turned his attention to Göttingen's records of distant earthquakes. From the study of seismic waves, researchers at the time knew that different layers of the earth were of different density. The outer layer, the earth's crust, shields the mantle, which, in turn, surrounds the central core. Little was known about the core's physical properties, however, or its size. Wiechert thought the core began at a depth of 1,500 kilometers. Gutenberg thought otherwise and began to calculate the travel times of earthquake waves below the 2,500-kilometer mark. Existing formulas did not work for the region Gutenberg wished to study; he himself had to calculate "the travel times of waves to be reflected and refracted at the surface of the core, outside as well as inside." He postulated different wave velocities in the core until the calculated travel times agreed with those observed on the seismograms. By "trial and error" (as he called it), Gutenberg found in 1912 that the boundary of the earth's core lies 2,900 kilometers below the surface of the earth. His figure still stands.

All young Germans were required to serve a year in the military. Gutenberg served his and then went to work in October 1913 as an assistant at the central office of the International Seismological Association, in Strasbourg, then a German city. When war came, Gutenberg worked as a meteorologist, traveling back and forth between the French and Russian fronts for several years; in his spare time, he continued to read and interpret the seismograms written on Strasbourg's instruments. When the war ended, Strasbourg became a French city, and the Allies banned German participation in international scientific organizations. This put Gutenberg out of a job, and he returned home in 1919, married a local girl, and went to work in his father's soap factory; but all the while he pushed forward relentlessly with his seismology research in the evenings.

New academic posts, along with meat and potatoes, all but vanished in inflation-ridden Germany. Soap, now a luxury, kept the Gutenberg family dressed and fed. "I know my boy's first shoes," his wife, Herta, said later, "we traded with the shoemaker." When circumstances improved, Gutenberg started lecturing without pay at the University of Frankfurt, and in 1926 he became extraordinary (associate) professor of geophysics there, continuing, however, to work part-time in the family business, which he inherited in 1927, following his father's death. "He was always working," Herta remembered. During this period, Gutenberg also supervised the university's seismological laboratory and turned out book after book, paper after paper, touching on all aspects of the earth, from the development of the continents to the temperature of the upper atmosphere.

The object of Gutenberg's ambition was Göttingen's vacant chair in geophysics. This was not to be. A well-placed colleague guardedly described the situation to him: "Now relationships, previously invisible, are becoming clear. . . . I'm pleased that you remain on the list, but don't be too disappointed if the anticipated succession . . . is envisioned in a manner different from what you expected." What this message probably signified may be gathered from the experience of Gutenberg's friend Theodore von Kármán, who had nursed a similar hope in the early 1920s, until he learned from Max Born that the number of Jews on Göttingen's science faculty was considered a matter of concern.

The difficulty of realizing his potential at home and the lure of a well-equipped Caltech laboratory made Gutenberg very receptive to Millikan's cable. He balked only at the salary offer of $5,000, which was $500 less than Wood's, and asked instead for $8,000 to $10,000. Millikan, who liked to boast that southern California's sunshine equaled several thousand dollars in salary, fumed. In asking for the higher salary, Gutenberg had figured in the cost of his two maids. Millikan advised the Gutenbergs to buy an electric iron, washing machine, vacuum cleaner, and other maid substitutes. "I think you will like the arrangement just as well, if not better, for it corresponds to a more just social situation, and in any case one is usually most content when he is living as his neighbors are living." Gutenberg had already determined that $5,000 was above the average professorial salary. When Millikan proposed $7,000—only five of the school's thirty-three full professors, including Millikan, earned more—he accepted. (Clinton Judy, the school's one-man English department, had a salary of $4,500.) In the autumn of 1930, Caltech's new professor of geophysics joined Wood and Richter in the two-story, red-tiled building, by then officially called the Seismological Laboratory.

Gutenberg's appointment decisively shifted the center of seismological research from Germany to the United States and the center of American seismology from Berkeley to Caltech. The number and breadth of the scientific papers Gutenberg, Richter, and Wood published alone or together in the *Bulletin of the Seismological Society of America* revitalized this single American journal in the field. Of the thirteen papers on the program at the society's 1931 meeting in Berkeley, six came from Pasadena and three bore Gutenberg's name. In truth, Gutenberg took up North American microseisms and the structure of the earth's crust in California with a vengeance. He turned to the laboratory's stock of seismic records and reanalyzed the data contained in seismograms of twenty-one southern California shocks. Between 1930 and 1932, he published two papers with Richter, one on the earth's mantle and the other on pseudo-seisms; with Wood, he collaborated on a study of blasting recorded in southern California; all three worked together on the 1930 Santa Monica Bay earthquake; and with Buwalda and Wood, Gutenberg tested seismological methods used in determining the earth's structure. "The earthquake business is picking up," Richter remarked to Wood some months after Gutenberg's arrival. Shock waves of a different kind soon reached the north.

Science is sometimes a back-scratching business. Co-workers in northern California had organized a regional seismological network in the San Francisco Bay Area along the lines of Wood's southland network. Over the years, an informal division of labor had grown up. It gave the Berkeley seismology professor Perry Byerly responsibility for tracking earthquakes originating north of Santa Barbara County. Wood in exchange had a free hand south of Mono Lake. Gutenberg tended to overlook the Wood-Byerly line. "This is another of the earthquakes which occur near our boundary line," Byerly teased Gutenberg on one occasion, "and which may lie in the territory of either. You have now had both the Coalinga earthquakes, which lie in my territory and this one, which is debatable. I therefore feel that I should have the opportunity of claiming the next quake which occurs in doubtful territory." But Byerly had no need to worry. The same attitude that caused Gutenberg to ignore the Wood-Byerly border made all of California too small for his studies.

Early in 1931, Wood began sending out a monthly bulletin that provided basic information about distant earthquakes recorded locally; he also listed the stronger local shocks, of which there were a great many. Ten years earlier, Wood had complained about lack of information about quakes in California; now, the files of his assistant Richter bulged with them. The volume of records made it all but impossible to distinguish between individual events. Wood knew better than to issue a local-earthquake bulletin under these circumstances. Better to wait until Richter could find a way to rank the tremors registered in southern California. "We needed something which would not be subject to misinterpretation in terms of the size and importance of the events," Richter later said. He solved the problem by inventing a new earthquake scale.

Scales available early in the twentieth century measured the relative intensity of shocks by the observed effects at a given place. It was significant, for example, whether chimneys or whole masonry structures fell down. The earliest widely used scale was the work of an Italian seismologist, M. S. de Rossi, in the late 1870s; F. A. Forel, a Swiss colleague, put forward a similar one in 1881; two years later, de Rossi and Forel agreed to cooperate, launching a seismology that flourished in Italy, as modification of de Rossi's scale proceeded apace. Between 1888 and 1904, G. Mercalli extended the scale from ten grades of intensity to twelve, while A. Cancani tried to connect the various intensity grades with ground acceleration. Some years later, German seismologists added refinements to the scale.

The Italian scales and their German improvements could give only a qualitative rating: the skill of the observer, the population

density in the shaken area, building materials, indigenous architectural tradition, and local geology all affected the registered
intensity. Acutely aware of these complications, Wood and
F. Neumann published in 1931 a modified version of the standard
scales for use in grading local shocks registered in Pasadena. A
shock of intensity seven, on a scale of twelve, indicated that people
had fled outdoors, chimneys had broken, and poorly built structures had suffered damage. An additional effect listed by them,
"noticed by persons driving motor-cars," added an up-to-date
touch. As Richter observed at the time, "In a region such as
Southern California, where a large proportion of the shocks occur
in almost unpopulated districts, while still others are submarine
in origin, any general procedure of this kind [using an intensity
scale] is out of the question."

The idea for a quantitative earthquake scale came to Richter
in the course of reading a paper by a Japanese seismologist,
K. Wadati. Wadati calculated the true ground motions of local
earthquakes from the trace amplitudes recorded by seismographs
at various meteorological observatories, and then graphed the
maximum motion at each station as a function of the distance
from epicenter to station. His curves could be used to estimate
earthquake damages near the epicenter, regardless of where the
shock occurred. What caught Richter's eye was that the curves
also allowed seismologists "to make a rough comparison between
the magnitudes of several strong shocks." Wadati referred to "the
magnitude of the earthquake"; his phrase and his method alerted
Richter, whose own search with Wood for a way to distinguish
large, small, and intermediate shocks had not been successful.
"Then I got hold of this . . . paper," Richter later told an interviewer, "and that gave me the idea of plotting up the data which
we had in a particular way, and it worked out much better than
I had expected and produced this definite numerical scale that
practically fell out of the data."

"Plotting up the data" brought Gutenberg into the picture.
Early in 1932, Richter had worked on a fresh group of local earth-

quakes recorded on Wood-Anderson instruments at the laboratory's seven stations. Unlike Wadati, he measured the maximum recorded amplitude directly from the seismograms, but ran into difficulty when he tried to plot the numbers along the vertical axis. Some of the amplitudes were as small as 0.1 millimeter, some as large as 12 centimeters. "Here is the problem, what do I do?" Richter asked the professor. If the two amplitudes were plotted on the same scale, either one would be too small to see, or the other would be off the page. "Try plotting them on the logarithmic scale," Gutenberg advised. When he did so, Richter saw that the curves connecting the scattered points for the different stations reading the same earthquakes lined up in roughly parallel formation on the graph. The magnitude scale all but stared him in the face. Despite Richter's account, some people may find it hard to believe that a theoretical physicist had to be told to use a log scale.

Richter took these curves, combined the essential characteristics of the group into a single curve, and assigned a "zero" magnitude to a shock that registers a maximum amplitude of .001 millimeter on a standard short-period torsion seismometer located 100 kilometers from the epicenter of the earthquake. An earthquake of magnitude 1 would register as 10 times bigger—.01 millimeter—on that seismometer, but it would also register 10 times bigger on all other seismometers at any distance. Magnitude 2 would be 100 times bigger, and so on. A magnitude 6 event is 1,000 times bigger than a 3. "If there was anything you could call an actual discovery that came out of that scale," its inventor once said, "it was that the biggest earthquakes were ever so much bigger than the little ones."

Richter tried out his method on the same group of local shocks that Gutenberg had studied in 1931 in connection with the travel times of earthquake waves in southern California. He analyzed twenty-one representative shocks involving distances from under 50 to over 500 kilometers, as "the most reliable test of the magnitude scale." In addition, he combed the files for examples of shocks with magnitudes between 0 and 2; Gutenberg's shocks,

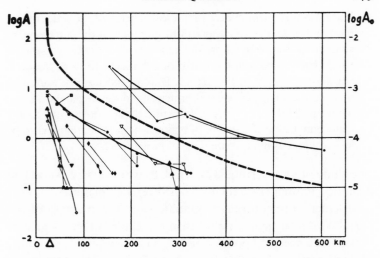

FIG. 1 Origin of the Richter scale plotting logA versus distance for various southern California earthquakes. A is the maximum needle deflection of a standard instrument measured in mm. The heavy dashed curve refers to the right-hand scale; for instance, at 100 km, $\log A_0 = -3$, at 300 km, $\log A_0 = -4$, and so on. The magnitude of the earthquake is $M = \log A - \log A_0$ for any given distance from the epicenter. For example, in one of the earthquakes shown, $\log A \approx -0.6$ at 300 km, meaning a needle deflection of less than 1 mm at that distance. Since $\log A_0 \approx -4$ at 300 km, the magnitude is $M \approx -0.6 - (-4) \approx 3.4$. (From C. F. Richter, *Elementary Seismology*, 341. Copyright © 1958 by W. H. Freeman and Company. All rights reserved.)

by comparison, all exceeded 3.5. Richter was able to show, partly by comparing the effects of shocks of various magnitudes by means of the modified Mercalli scale of 1931, that the torsion seismometers installed in southern California registered all the earthquakes "likely to be felt" within the 500-kilometer radius of the Pasadena program.

The statistics Richter compiled in the course of applying the magnitude scale undermined Wood's hypothesis that small quakes act as precursors of large ones:

For a shock of magnitude 4.5 represents energy of the order of one thousand times that of a shock of magnitude 3; if the latter were to account for the same release of strain in a given time, such shocks would have to be of the order of one thousand times as frequent, which is far from being the case. . . . Thus the conclusion is warranted that seismic energy is released principally in the larger shocks . . . while smaller shocks, occurring from time to time, do not appreciably contribute to the adjustment of regional strain, but are rather to be looked upon as minor symptoms of its existence.

The study of local earthquakes, a task to which Harry Wood had dedicated himself since the teens, had reached maturity.

Richter's instrumental magnitude scale is a measure of the overall size of a quake, rather than a measure of the degree of shaking at a specified point. The magnitude number is the same, regardless of the location of the recording stations. Outside of California, Richter's scale was tested extensively in New Zealand; prompted by the destructive 1931 Hawke Bay earthquake, the government installed the Wood-Anderson seismograph and other short-period instruments at several key stations. Seismologists there assigned earthquake magnitudes to local shocks by means of Richter's tables for California shocks. Gutenberg then played a major role in extending Richter's local scale to earthquakes anywhere. Wide acceptance of the scale came later on, around 1950, when stations in different countries started assigning magnitude numbers to the shocks they reported. And now "Richter scale" is virtually a household phrase.

Besides working on local earthquakes and the magnitude scale, Richter began a sustained scientific collaboration with Gutenberg that lasted until Gutenberg's death in 1960. Gutenberg had his own research style. "He would begin by outlining a comparatively limited investigation based on material in hand or readily obtainable," Richter said later. "As work continued, more and more data of different kinds would be drawn in, and the scope of investigation enlarged, resulting either in a larger publication than originally proposed, or in a series of papers covering different points." A good example is the investigation of three large 1931 earthquakes, two in Sumatra and one in the Solomon Islands. Excellent records of these shocks existed, and before long Gutenberg and Richter knew the times of occurrence and epicenters with much greater precision than is usually the case. The good data inspired Gutenberg's next move. He and Richter expanded their study to include 108 additional earthquakes and, by 1934, had embarked on a thorough revision of the standard time-distance tables used in calculating the velocity of seismic waves. They completed their study of seismic waves shortly before World War II. Richter described to Harry Wood what he and Gutenberg had accomplished as follows: "The whole group of data makes possible a complete revision of the results by seismic methods on the structure of the earth, both interior and crustal."

When Buwalda learned in 1934 that Carnegie's Merriam had decided "to withdraw gradually" from the seismology program, he urged Millikan to save the scientific partnership. He knew that the Carnegie Institution's involvement went counter to its policy. "If we had merely continued to furnish housing as we did during the first few years," Buwalda wrote, "the program would in all probability [have gone] forward as a Carnegie program." In other words, if the institution paid the bills, it expected to control the laboratory program; Day had said as much to Wood during the great stationery debate. Millikan insisted that the laboratory's accumulated knowledge belonged to the world, not simply to the local community; Merriam countered that only Caltech—and Caltech money—could make the study of California earthquakes respectable. After protracted negotiations, Millikan got what he wanted: the Carnegie Institution agreed to continue its 1935 level of spending through 1938. On January 1, 1937, Caltech took over the administration of the laboratory; Charles Richter and Hugo Benioff were appointed assistant professors of seismology at the Institute. By the end of 1941, the Carnegie's financial involvement in the seismological program had come to an end.

Only Wood remained on Merriam's payroll, but his days in

the Seismological Laboratory were numbered. In 1934, a virus attacked his spinal cord, making it difficult for him to speak, eat, and walk. Although he recovered sufficiently in 1937 to return to work from time to time, Wood never again played an active role in the laboratory. The true cost of his illness, he told Buwalda in 1938, could be measured in the laboratory's relentless emphasis on distant earthquakes. "Richter would have worked much more with me and less with Gutenberg, the final Long Beach earthquake paper would have been completed and published, and other local earthquake papers as well," had things remained as they were. If the laboratory's direction is not corrected, Wood complained, "my own work and effort will have been wasted largely—for with a very small fraction of the energy I have expended I could . . . have developed a first class teleseismic institute at far less cost." Between bouts of illness, Wood worked up a number of papers for publication, including several on historical shocks. He died in Pasadena in 1958, two years before Gutenberg, at the age of seventy-eight.

Buwalda sounded a triumphant note in 1941 when he contrasted Wood's approach with that of Gutenberg, Richter, and others at the laboratory. In his first formal report to Vannevar Bush, Merriam's successor at the Carnegie Institution of Washington, Buwalda described the efforts of Caltech's scientists to understand deep-focus earthquakes, the global distribution of earthquakes, and the causes of shocks, among other problems:

> As you know, seismology has during the past two decades become the most powerful tool ever developed for exploring the interior of the Earth. Most of what is now known about the structure both of the crust of the earth and of its deeper shelled portions has been learned from the transmission, reflection, or refraction of earthquake waves. . . . The earth structure problems being investigated are . . . among the largest and most important problems dealing with the earth under attack anywhere. This is a new field being investigated by new methods. . . . The problems are not merely California problems, but they are of worldwide interest.

EIGHT

Aeronautics and the Airplane Industry

The development of the flying machine from the rather primitive contraption of the Wright brothers to the complex and efficient high-speed airplane of today has been most spectacular. Yet when I fly in bumpy weather or when I am forced to wait hours at an airport because of the weather—or because of the ignorance of the weatherman—I wonder whether our achievement is really so miraculous.

—THEODORE VON
KÁRMÁN, 1954

MOVIES AND REAL ESTATE were southern California's biggest industries in the 1920s, but aviation was catching up. Lured by the prospect of 350 flying days per year, a climate that allowed year-round testing of open-cockpit airplanes, and the possibility of assembling and parking planes outside, the airplane makers had begun to flock to the southland. Donald Douglas, for example, worked for one of the earliest airplane producers, the Glenn L. Martin Company, first in Los Angeles and later in

Cleveland, before he founded his own aircraft company, in 1920, in the back room of a Santa Monica barbershop. Others followed, including John K. Northrop and Alan Lockheed, who began building airplanes nearby in 1927. By then, the locally based aircraft industry was contributing $5 million a year to the region's economy.

The world's first powered airplane, which the Wright brothers designed, built, and flew at Kill Devil Hills, near Kitty Hawk, North Carolina, on December 17, 1903, was made of wood, wire, and cloth. A "birdcage with all the structure visible" is how one aeronautical engineer later described the Wright brothers' *Flyer*. Before flying their machine under power, they had flown it as a glider; before that, as a kite. They had also tested wing shapes in their own small wind tunnel, performing close to a thousand tests. This sort of careful, step-by-step development led the Wrights to the idea of wing warping (twisting the wings slightly) and a movable vertical tail. Both became integral features of their airplane. Then they added propellers, two of them, which they built themselves, and put a motor on their airplane, a twelve-horsepower engine. On its inaugural flight, the *Flyer* took off under its own power and flew 120 feet, in the space of twelve seconds. On the fourth and final flight of that day, it flew 582 feet, staying in the air for fifty-nine seconds. "They did it! They did it! Damned if they didn't fly!" one eyewitness later reported in the town of Kitty Hawk.

But the public, including the press, wasn't inclined to believe the news. An editorial announcement conceding "the epoch-making invention of the first successful aeroplane flying machine" did not appear in the pages of the *Scientific American* until December 15, 1906, three years after the Kitty Hawk flights. In the meantime, the Wrights had continued testing and improving their new machine. After much hesitation, in 1908 the Army Signal Corps ordered one of them, at a purchase price of $25,000. Specifications called for it to go 40 miles an hour, to hold fuel to last 125 miles, and to seat two people—a pilot and one passenger. A year later, Orville Wright successfully tested such an aircraft for the War Department at Fort Myer, Virginia.

Even so, neither the army nor the general public took this flying business very seriously. In an effort to win support for aviation, the Wrights promoted air shows and air meets; and when spectators wearied of aerial acrobatics and stunt flying, aviation buffs looked for more dramatic feats. In 1911, Calbraith Rodgers flew a modified Wright plane, named the *Vin Fizz*, from Brooklyn, New York, to Pasadena, landing in a park just across the street from Throop College. This first transcontinental flight took Rodgers forty-nine days in the air and more than two months' travel time, including one crack-up along the way, to complete. By then, some dozen established American aviation companies were building airplanes, although the total number of planes sold up until then in the United States was not very high, perhaps two hundred.

When Europe went to war in 1914, Germany had 230 airplanes at its command, England 110, and France a few more. The United States had just 6 airplanes in its aviation arsenal. Nor did the picture change quickly. When the United States entered the war in 1917, the armed services still had fewer airplanes than Germany had entered the war with. Backed by $640 million in congressional appropriations, the Signal Corps placed contracts for more than 300 planes of wide-ranging design. As one aviation historian has noted, "The orders were placed with 16 firms, only a half-dozen of which had ever produced as many as 10 planes or more. None of the companies had ever produced anything more advanced than a training plane, and there were no more than 10 designers capable of that job." In the final year of the war, however, American production exceeded 14,000 airplanes, a record number.

On other fronts, progress was less evident. America went into the war lagging behind Germany in airplane design and construction and never did catch up. Hugo Junkers, a professor of mechanical engineering at Aachen, had in 1917 built an all-metal monoplane, one with thick wings, internally braced, and strutless. The thick cantilevered wing had been designed with the help of a theoretically minded professor of aeronautics, Theodore von Kármán. America had no Junkers and no von Kármán. Advances of this kind did not occur over here. In the immediate postwar period, American aviation manufacturers continued, by and large, to build canvas biplanes with thin wings, held together with cables and struts.

For historical reasons still only partially understood today, the Wrights' invention of the airplane did not foster the study of aeronautics as an exact science in the United States. Rather, Germany, followed by France and England, Italy, and the Soviet Union, took the lead in establishing the science of flight as an academic discipline. While U.S. aviation pioneers remained resolutely empirical, aeronautical research and development at the University of Göttingen, in particular, early stressed theoretical as well as experimental research.

In this time, however, the United States was not willing to accept second-class standing for long. In 1926, Daniel Guggenheim set up a $2.5 million fund to start seven aeronautical schools in the United States, including one at Caltech. To lead the Pasadena program Robert Millikan went after none other than Theodore von Kármán, Göttingen Ph.D., class of '08, and head of the Aerodynamics Institute at the Technical University of Aachen. By the mid-1920s, this accomplished physicist, aerodynamicist, and applied mathematician had built up an aeronautics establishment at Aachen second in importance only to Ludwig Prandtl's at Göttingen. Von Kármán's association with Caltech would figure prominently in the rise of American aeronautics in the 1930s.

The roots of aeronautics at Caltech, however, go all the way back to 1917. As America prepared to join the war against Germany, the trustee George Ellery Hale promoted aeronautical research as a way for the school to gain national stature. Most important in this area was the arrival of Harry Bateman, an English mathematical physicist, and Albert A. Merrill, an American inventor.

Upon learning in 1916 that Throop College was scouting for "a mathematical physicist to help in some theoretical investigations," Bateman had contacted Hale. Despite his modest rank as a lecturer in applied mathematics at Johns Hopkins University, in Baltimore, the thirty-three-year-old mathematician had impeccable academic credentials. Following his graduation in 1905 from Cambridge University, where he distinguished himself as winner of the Senior Wrangler and the Smith's prizes, Bateman studied at Göttingen and in Paris. Returning to England, he taught first at the University of Liverpool and later at the University of Manchester, before immigrating in 1910 to the United States.

By 1916, Bateman had compiled an impressive record as a mathematician. He had to his credit some seventy scientific papers, on topics ranging from geometry to earthquake waves; a British Association report on the history and theory of integral equations; a Cambridge University monograph on Maxwell's equations; and a textbook in press on differential equations. But tenured academic jobs in mathematical physics were just as scarce in the New World as in England. Bateman had barely gained a toehold on the academic ladder at Hopkins, where he had also earned a doctorate in 1913. To make ends meet, he lectured at the Bureau of Standards and reviewed papers for the Weather Bureau. In his letter to Hale, Bateman pointed out how much he would "welcome an appointment which enabled [him] to devote the whole of [his] time to useful investigations."

Fresh from lecturing in Pasadena, Robert Millikan (the college's new director of physical research) was dispatched from Chicago to Baltimore in February 1917 to interview Bateman. Millikan came away impressed, and suggested to Hale not only that Throop take Bateman but also that, "in addition to keeping with the aeronautics problems," Bateman "be tried out on one class . . . [such as] calculus." From their conversation together, Millikan gleaned other useful information. "I found that $1800 would get him and I think he ought to get that much in view of his age and accomplishments."

Hale had lines out everywhere. Around the same time, when Hale told MIT's E. B. Wilson, the author of a textbook on aeronautics, about Throop's plans to build a wind tunnel, Wilson chided him for putting one up, citing the "great deficiency of brains to handle what experiments the existing wind tunnels could make." In defending the construction, Hale had hinted at special circumstances; Wilson, determined to involve someone trained in mathematical formulations in the project, replied, "If you could hire Mr. Bateman who used to be at Johns Hopkins and who is a very able man, who looks as if he were running into consumption . . . you would have a great addition to your school." A month later, in the spring of 1917, President Scherer of Throop offered the young theoretician a position at the school.

Bateman was not impressed. The president's letter had omitted a number of details. What would his title be? Was it a permanent job? Would he work alone or with colleagues? What were the teaching load and the class size at Throop? Was he free to work on problems of his own choosing? How much time would the laboratory work entail? Could he count on getting a salary raise fairly soon? Scherer cabled him the good news: "Position will be Professor of Aeronautical Research and Mathematical Physics." The cable continued,

> Position permanent. Independent of war developments. No routine undergraduate instruction expected, but advanced courses for few students may be offered at your pleasure. You will be independent in your work. We want you primarily to develop theoretical side of aeronautics, also co-operate in suggesting experimental researches. Will contribute toward travelling expenses. Living cost considerably less here than in some eastern cities.

The part about the cost of living in southern California struck a sour note. If they meant to make him a professor, Bateman told Millikan by return mail, "I think I should also be offered a Professor's salary." He suggested $3,600, twice what Scherer had offered, and twice what he was making at Hopkins. None of

Throop's eight full professors then earned more than $2,400, and faculty salaries started at $1,100, not encouraging news. "I doubt the wisdom of going above 2500 if it is possible," Millikan scrawled across the top of one letter routed to Scherer. Bateman finally accepted $2,000, plus $500 for traveling expenses.

Aeronautical research could scarcely be done without a wind tunnel. To provide funds for it, Hale turned to the Pasadena millionaire Tod Ford, Jr., the son of a Youngstown steel industrialist. Ford's passion for aviation would be converted into an aeronautics laboratory. A dedicated flier before the war, Ford had joined the Lafayette Squadron, the American volunteers in the French air service. Grounded by an ear injury, he returned to the United States and took up residence in Washington for the duration of the war, promoting the work of the National Research Council. By the end of 1916, Hale had sold Tod Ford on the laboratory idea.

Hale had already dangled Bateman's name in front of the army's chief of aviation, Colonel George Squier, in the hope of getting "important research" for the wind tunnel. In reply to Millikan's initial report on Bateman, Hale had recommended that he be hired as soon as Throop had Squier's word that there were problems in aeronautics that the mathematician could profitably work on. As head of the Signal Corps' aviation section, Squier was a member of Millikan's NRC physics committee, and they worked together closely. Indeed, he later persuaded Millikan to accept a major's commission in the Signal Corps. Squier is reported to have offered "advice" on the establishment of the aeronautical laboratory.

But when Hale offered to get someone from the military to teach—without pay—an aviation course at the college, the board chairman Arthur Fleming balked. "We must beware of seeming to go too deep into military work." Surprised perhaps by this "somewhat adverse reaction," President Scherer quickly pointed out there would be no actual flying and no significant military applications, only airplane theory and design.

In June 1917, the school's trustees authorized construction of a wind tunnel for aviation experiments. A press release the following month explained that Bateman was expected to "devote his entire time to the work of the College in connection with aeronautics research utilizing as his laboratory the wind tunnel for which the contract has just been let." The local paper reported that his work would "relate to the stability of airplanes during flight." Such drumbeating was a far cry from the attitude of Bateman, who had once downplayed his chances of finding employment "with a physical laboratory because [he had] not done much experimental work."

Still, the aeronautical laboratory did not progress remotely as fast or as smoothly as its promoters had anticipated. Ford was in France and Scherer in Washington. "Bateman, of course, is reading and giving a few lectures," one school official close to the project remarked,

> . . . but the wind tunnel, complete except for the balance and propeller, is doing nothing. Even after we get the balance and propeller it will take a little time to get them installed, get the electric wiring done, and, as Tod Ford says, to get the tunnel "tuned up." It seems that each tunnel has to be individually adjusted so that the air currents flow through it in proper lines and without eddies.

In the end, the wind tunnel, a square wooden tube measuring four feet by four feet and fifty-two feet long, built to test the effects of air currents ranging from ten to forty miles an hour on model planes, took more than a year to complete.

It was formally inaugurated in August 1918, three months before the war ended, and three months after A. A. Merrill joined the enterprise. Built primarily for war service, the college's wind tunnel—southern California's first—of course served no current military use.

Aeronautics had become, nonetheless, a permanent course at the college. What Bateman may have lacked on the experimental side of this branch of science, Albert Adams Merrill more than made up for. A founding member of the Boston Aeronautical Society in 1894, Merrill was performing gliding experiments and publishing papers on airfoil design before the turn of the century. He gave his first aviation speech in 1892, when graduating from English High School in Boston; he learned to fly in 1911 at the Wright brothers' airfield in Dayton, Ohio, and lectured in 1913 on aviation at MIT. A trained accountant, Merrill was working at Price Waterhouse in Los Angeles in 1918 when he was hired as Bateman's assistant, becoming Caltech's first teacher of aeronautics. From the first, Merrill operated the small wind tunnel; by 1921, he had been promoted to "Instructor in Experimental Aeronautics and in Accounting." He taught airplane design, a course that grew out of his own experiences in constructing and rigging aircraft. And his laboratory was the wind tunnel. His style of teaching was simple and straightforward. "I can take seniors into the laboratory," he once admitted, "and show them practically how we handle research problems, but if I should attempt to handle the engineering end I would fail. . . ." After he had the wind tunnel up and running, Merrill began work on a plane design featuring a movable wing.

Bateman, for his part, specialized in finding particular solutions to complicated equations used by physicists and applied mathematicians. And the two men were different in other ways: Bateman was as shy and unassuming as Merrill was brash. Indeed, his Caltech colleague E. T. Bell, fearful that Bateman might undermine his chances of election to the National Academy of Sciences by listing too little on his curriculum vitae, counseled, "Spread yourself; it pays, in our glorious country, to kick over the bushel and let your light to shine before men that they may see your good works. . . ."

Bateman was the theoretician, Merrill the tinkerer. Aviation had always belonged to the amateurs, and Merrill stood high in their ranks. A self-taught inventor well versed in the practical side of aeronautics, Merrill had the field to himself at Caltech for a while. However, late in the 1920s, Arthur E. Raymond, a member of the technical staff of the Douglas Aircraft Company and an expert in designing planes, was hired to teach a class in aircraft design. By then, the day of the amateur was drawing to a close.

In his initial appeal to Harry Guggenheim, at the end of 1925, for funds to establish a research center at Caltech "for advancing the science and art of aeronautics," Robert Millikan leaned heavily toward fundamental aeronautical research. When Guggenheim balked at the heavy emphasis on "research" and plumped for more applied activity, the pragmatic Millikan tailored his argument to suit the sponsor. As he wrote Guggenheim, "there is a very definite practical side, not only to our program but to our actual accomplishment in aeronautics at the Institute," and he went on to describe Merrill's experiments with gliders and a plane Merrill had designed.

Millikan's letter met with some success, because Guggenheim replied that he found what had been said in regard to Merrill's plane "most interesting." "If on my return [from a tour of European aeronautics centers] it would seem to you," he wrote Millikan, "that this matter is one in which the fund [the Daniel Guggenheim Fund for the Promotion of Aeronautics] could be of assistance, I shall be very pleased to see that the matter has every consideration."

Even as he used Merrill's accomplishments to woo Harry and Daniel Guggenheim (Harry's father), Millikan had begun to reorient the Guggenheims toward his own way of thinking. Thus, in a "condensed reformulation" of his original proposal for a school of aeronautics at Caltech, Millikan asked for an endowment for "the support of an outstanding man of Bateman's type and an assistant." Guggenheim ultimately invited Millikan to submit his budget proposals to the fund's executive committee, because the fund's board of directors had approved "an appropriation of funds for the establishment of a school of aeronautics at the Institute, in accordance with the general purposes of your proposals." They provided the money for a building, equipment, salaries for ten years, research funds for five years.

In describing the new aeronautics program, Caltech officials moved one step closer to shoring up the practical side of the enterprise. "The facilities of the Douglas Company in Santa Monica," one brochure noted, "with its large corps of engineers will supplement those of the Institute for both instruction and research purposes. . . ."

Although Millikan had asked the Guggenheim Fund for support for a second theoretician, Bateman and the physicist Paul Epstein urged Millikan to select an aeronautics adviser experienced on the experimental side. When Millikan called for suggestions, Epstein volunteered the name of an old friend from his student days in Germany, the Hungarian-born engineer and applied scientist Theodore von Kármán.

Von Kármán was an aerodynamicist, trained in physics and mathematics, and he believed strongly that engineers should be taught to use mathematics to solve physics problems. Without question, von Kármán had his feet planted in both worlds. He had done work on the buckling of columns, on the stability of vortex patterns that form behind stationary bodies in flowing fluids, and, with Max Born, on the lattice dynamics and vibrational frequencies of crystals, advancing work done earlier by Albert Einstein and Peter Debye on the heat capacity of solids. He brought a mathematically sophisticated point of view to all of these problems.

At Millikan's direction, in July 1926, Epstein wrote von Kármán at Aachen (the letter caught up with the vacationing aeronautical engineer in Belgium) inviting him to visit Pasadena that fall. Epstein thought von Kármán could give a few lectures and look over the construction plans for a new wind tunnel and Merrill's designs for a new plane. The question of whether to send Caltech's students abroad to gain experimental experience, or whether to create a full-time position locally for a "well-trained foreign brain," also had to be decided.

Von Kármán visited Caltech in 1926 under the auspices of the Daniel Guggenheim Fund for the Promotion of Aeronautics. He returned in 1928 for an exchange semester (Epstein went to Aachen); he joined the school as a research associate in aeronautics in 1929; and he was appointed professor of aeronautics in 1930, and director of the Daniel Guggenheim Aeronautical Laboratory (GALCIT).

By his own account, politics, not science, brought von Kármán to America in 1930. He disliked not only the growing German militarism but also the increasing number of anti-Semitic incidents in Aachen. Letters of von Kármán's suggest that a personal encounter with anti-Semitism as far back as 1922 left him ripe for Millikan's offer.

The episode involved his physicist friend and scientific collaborator Max Born and the mathematician Richard Courant. They had proposed von Kármán as Ludwig Prandtl's successor, following rumors that Prandtl might leave Göttingen for Munich. Their suggestion came to nothing, because Born and Courant, also both Jewish, were unwilling, "to fight it out for you against the enemies of Israel," as Born wrote to von Kármán. "The prospect made me miserable," Born added. "I simply did not have the physical and emotional strength to unequivocally pursue this goal. . . . I wanted peace and quiet. That wasn't nice of me." As it turned out, Prandtl did not leave, but the fact that no overture was made to him, because of his Jewishness, could not have failed to leave its mark on von Kármán.

On his visit to the Pasadena campus in 1926, von Kármán saw—and disapproved of—the plans for a new, ten-foot-diameter wind tunnel (the original Throop tunnel had burned down in the early 1920s). This wind tunnel, he felt, was a step in the wrong direction.

The wind tunnel is the primary experimental tool of the aerodynamicist. In it, scale models of newly designed aircraft can be tested for lift, drag, and stability before the neck of a test pilot is risked. In his autobiography, von Kármán recalls that, on the spot, he used the back of an envelope to sketch a closed circuit type of tunnel in which the air could still freely circulate. He felt that with the shape he had drawn, it would be possible to increase the efficiency of Caltech's tunnel. The Soviets, von Kármán further relates, "had just announced the successful construction of a wind tunnel with an energy factor between 5:1 and 6:1." The higher the energy factor, the more energy efficient the tunnel. When von Kármán looked at Caltech's plans for an open-return tunnel, which involved taking in air and exhausting it, he knew that its energy factor would not exceed 3:1. The closed air-circulation system could do better. To build it, he enlisted Robert Millikan's son, Clark Millikan, one of Harry Bateman's graduate students, and a fresh new Caltech Ph.D. in physics, Arthur L. ("Maj") Klein.

Klein had stumbled into aeronautics. He'd entered Throop as a freshman in 1916, majoring in mechanical engineering, but had later switched to physics. In his oral memoirs, Klein says he changed majors to delay going into the family's retail business ("I didn't want to sell socks, shirts, and underwear"). Besides having a natural gift for seeing how things worked (he'd been kicked out of Caltech once for pranks ranging from rewiring the electrical laboratories to making the air blow the wrong way in Merrill's wind tunnel), Klein wanted to stay around the Institute, and he didn't need to find a steady job—he had a regular income from the family business. After taking his degree under Robert Millikan in 1925, he became a research associate, occasionally teaching a class. Mostly, he helped Clark Millikan and others who were working down the hall in Bridge Laboratory. "They were hopeless, mechanically," he later recalled. "They were building an airplane, and they were having troubles with that. . . . They did the stress analysis and aerodynamics very well, but just how you put things together, they didn't know."

When the construction of the Guggenheim wind tunnel got under way in earnest, in 1928, Klein found his true vocation. He remarked many years later, "Everybody in the aeronautics de-

partment [during the summer] went on vacation. Then the fore-
man came over and said 'How do you do this?' and 'How do
you do that?' and I would tell him, and nobody seemed to object
so I went on telling him." Since he was consulted so often, the
issue of which department he belonged to arose. Klein recalled,
"They decided I'd better be put in the Aeronautics Department,
because somebody might question what I was telling people to
do, and seeing that I was the only one around to tell them, why
they'd better give me a little authority. So, that's how I got into
the Aeronautics Department." People called him Maj because he'd
been a cadet major at the school, and the name stuck.

Maj Klein designed much of the wind tunnel. He designed
the balances used to record the various forces on a model plane,
the wind tunnel rigging, and the water-cooled vanes, among other
mechanical features. Besides cooling the tunnel, the vanes also
helped keep the air flow in the tunnel smooth and uniform. The
wind speed inside the tunnel is controlled by a fan, which blows
a uniform current of air through the tunnel and over the model.
The model itself is made from a hardwood, lacquered and polished
to a high shine, and suspended inside the wind tunnel. When
Caltech's ten-foot-diameter wind tunnel was first put to the test,
in 1929, Klein and Clark Millikan were sure something was wrong
with their experimental data. "Don't yell!" Klein twitted a fellow
engineer at Stanford. "Our energy ratio is somewhere between 5
and 6. I know you won't believe it; we can hardly believe it
ourselves." The energy ratio turned out to be an impressive 5.6
to 1. "In fact it was so much better than the Caltech people had
expected, that they had not designed the propellers large enough,"
von Kármán later remembered. "They were forced to redesign
the propellers to take full advantage of the speed possibilities of
the tunnel." At top speed, in fact, the stream of air in the tunnel
reached over two hundred miles per hour. The entire aeronautics
department at Stanford came down to see the tunnel in operation.

Robert Millikan had hoped, among other things, that von
Kármán would review Merrill's movable-wing tailless biplane dur-

ing his first visit to Caltech in 1926. Clark Millikan had already recommended the plane, without reservations. "The pilot can take off and land by merely shifting his wings and without touching the elevator," Clark told his father, after watching the plane fly at Ross Field, Arcadia, in the spring of 1926. Its commercial future seemed assured. "The machine can be trimmed so as to fly with hands off the control at any speed, throttle opening, or rate of climb or descent. . . . The practical advantages seem very considerable: chief among them of course being safety, as it will be impossible to stall the ship involuntarily even in a fog." A strong endorsement indeed. But the elder Millikan's hopes may have been tempered, for von Kármán apparently had his reservations. Later he described Merrill's idea "to control flight by turning the entire wing, instead of a contral flap," as "one of those ideas which strike you by their simplicity, but never lead to success due to practical difficulties."

Criticism of Merrill's cherished design led to bruised feelings. In November 1926, Robert Millikan issued a "Memorandum" on the "Rules of procedure which are universal at the Institute." One of these emphasized the cooperative nature of the work, in which "determinative judgments are joint judgments arrived at by the method of conference." Merrill was the memorandum's sole target. In it, Caltech's head promised "to push the development of Mr. Merrill's idea just as fast as possible in the light of its present and future performance," and though its inventor would continue to retain the rights to all patents, the Institute nevertheless wanted more say in decisions about his plane. Merrill apparently agreed to the provisions of the memorandum; von Kármán's visit may have precipitated it. The memorandum, for its part, marked a turning point in the fortunes of Merrill at Caltech, for the fiercely independent inventor found, much to his dismay, the biplane to be less and less his private domain and more and more in the hands of others in the years thereafter.

At any rate, Merrill's experimental plane was damaged in a fire on the Caltech campus in the summer of 1927. It was replaced by the C.I.T. 9, a little green two-seater, designed and built by Maj Klein, Clark Millikan, and Merrill working together. Better known among the students as "the Dill Pickle," Merrill's new plane underwent initial testing the following May, flying a distance of thirty miles. Soon afterward, the plane was demonstrated in the Mines Field air meet in Los Angeles, but disaster struck in the fall of 1928. The Dill Pickle took off, hit a haystack, and crashed. As Klein, who designed the plane's hardware, remembers it, the plane "ground looped and broke the wing. And that was the end of it." The pilot was not hurt, but Merrill's feelings were more than bruised. He blamed Clark Millikan for the plane's crash. "Of course any ship is liable to ground loop," Merrill wrote to him, "but the question is, does a particular design increase the danger from a ground loop?" He answered his own question: "I contend your design does." The 1928 crash of Merrill's biplane may have hastened his departure. In any event, Merrill left Caltech at the end of the year.

In 1928, Robert Millikan approached Donald Douglas about the possibility of hiring an instructor to take over the practical work in aircraft design. Douglas suggested his assistant chief engineer, Arthur Raymond. A graduate of Harvard, Raymond had studied aeronautics at MIT before joining Donald Douglas's company in 1925. In September 1928, Raymond started teaching a class in airplane design at Caltech one day a week, usually Saturday morning. His first class ("an Elementary Class," according to Raymond) consisted of Bateman, Merrill, Clark Millikan, Maj Klein, and Ernest Sechler, a structural engineer. Millikan had just earned his Ph.D.; Sechler was a graduate student. The course, thirty weeks in length, covered such topics as design loads, drag, thrust, weights, and tail-surface analysis—in brief, the forces that are exerted on an airplane.

Raymond taught the same elementary class for six years, until June 1934, when he was appointed chief engineer at Douglas. But the course went on, picked up by other Douglas engineers. After that first year, Raymond also offered an advanced class. He assigned homework, which Klein supervised. "This was supposed to take four hours," Raymond later said, "but actually took much longer, according to the students, who considered me a hard taskmaster." Many of the graduates of that course went to work for Douglas, including W. Bailey ("Ozzie") Oswald, the Guggenheim Aeronautical Laboratory's first Ph.D. Indeed, records from that laboratory reveal that of the thirty most prominent graduates in the 1930s, nearly half—those who were theoretically oriented—joined universities, and the others worked in industry, especially the local aircraft companies. In general, students who did their work in aerodynamics (like Oswald) went into the aircraft industry, and those who specialized in fluid mechanics (studying problems such as turbulence and the boundary layer) became academics.

For his thesis, Oswald tackled the problem of working out an improved method of performance analysis by means of a slide rule. At that time, aircraft performance was measured by tedious graphical methods. Raymond, who suggested the problem to Oswald, wanted a faster method of gauging such performance criteria as gross weight, engine power, endurance, and cruising speed. Oswald's general formulas and charts for the calculation of the behavior of an airplane became one of the government's best-selling technical reports. Once Oswald had earned his Ph.D. degree from Caltech, in 1932, he was immediately hired by Raymond as Douglas's chief aerodynamicist, the company's first. Its engineers were then in the process of gearing up to begin work on the Douglas DC-1 (for Douglas Commercial) for Transcontinental and Western Air. The specifications for the new plane, Raymond later wrote, filled "15 legal-size pages. By then I was sure we were faced with a challenge, so about the first action I took was to hire Ozzie, temporarily, I told him, for two weeks. . . ." The two weeks turned into a permanent job, which Oswald held until he retired.

Raymond also recruited Sechler, another graduate of his Saturday morning class in Pasadena. At the suggestion of his instructor, Sechler had turned himself into an expert on sheet metal structure. Douglas, in turn, supplied the test samples. For the DC-1, Sechler worked out design criteria for stressed-skin construction. In particular, he used the machines in the Caltech structures laboratory to establish a series of performance curves. These curves, critical in the design of airplanes, gauged the loads that metal could take before breaking, depending on such factors as the thickness of an all-metal wing or fuselage, the amount of stiffener, and the variance in the stiffener spacings.

When TWA ordered the DC-1 from Douglas Aircraft in 1932, it specified a type of construction that had been pioneered by Northrop. This was a comfortable connection for Caltech. Wind tunnel work on the Northrop Alpha—a low-winged monoplane that preceded the DC-1—had been done at Caltech under the supervision of von Kármán, Clark Millikan, and Maj Klein. This Northrop Alpha design testing, a very early trial for the Guggenheim Laboratory tunnel, had been tense. Buffeted by the tunnel's 200-mile-an-hour winds, the Northrop Alpha's tail began to shake violently, making controlled flight impossible. Klein and Millikan at length traced the problem to turbulence at the juncture of the wing and fuselage. In his autobiography, von Kármán explains the vibrations: "A sharp corner where the two came together caused the air to decelerate as it swept past and to form eddies. As these eddies broke from the trailing edge of the wing, they hit the tail, the stabilizer, and the attitude control, causing them to vibrate." The solution rested on an appreciation of a complex hydrodynamic phenomenon known as vortex shedding, first explained by von Kármán himself in 1911. Caltech's aeronautical engineers had solved the problem by placing a smoothly shaped metal device, a small aerodynamic surface called a fillet, between the wing and the fuselage; that smoothed the air flow and stabilized the ship's tail.

As it turned out, French airplane designers in the early 1930s were experiencing the same problem. Von Kármán, in Paris in 1932 to give a lecture, described the steps Caltech engineers had taken to stop the model's vibrations. "One of the prominent designers told me later," he subsequently related, "that after my lecture he tried a fillet right away on his new prototype and had success. Thus in France the fillet was connected with my name and was called a 'karman.' The French say an airplane has a 'big karman' or a 'small karman.'" In fact, as von Kármán often pointed out, "The invention of the wing-fuselage fillet was in reality a joint work of our C.I.T. team." If anything, he said, it should have been called the Klein fillet.

The lessons learned in testing the Northrop model were soon applied to the Douglas DC-1. In his autobiography, von Kármán writes, "I enjoyed climbing into the ten-foot wind tunnel with a wad of putty, and imagining myself being the airplane I tried to feel where I might be pressed by an element of air." In time, two hundred test runs of the Douglas 1:11 scale model plane were made in Caltech's wind tunnel during the fall and winter of 1932. Descriptions of the tunnel in that era convey a sense of order and efficiency, even a touch of science fantasy. "The model is suspended from seven sensitive balances on the floor above the wind tunnel," begins one such description,

> and these record all the forces and moments which the wind exerts on the model. These balances automatically balance themselves and indicate exactly the loads carried by the model at any attitude. The control of the entire system is accomplished by one man at a control table where are collected instruments for observing and controlling the wind speed, for setting the model at definite attitudes, and for controlling the balances. The experimental data are recorded by a second worker and are then taken to the computing room where they are reduced and finally plotted in the form of curves which show the aerodynamic characteristics of the model.

In operation, the wind tunnel, of course, was noisy and its insides strictly off-limits to workers.

The DC-1, an all-metal prototype, had its first flight in July 1933, less than a year after Raymond had laid out the first sketch.

This craft, the only DC-1 ever built, was still flying in 1939. The DC-2, Douglas's first production model, of which 130 were built, followed. Then, in 1935, the company produced the airplane that turned flying into transportation: the 21-passenger, twin-engine DC-3. By 1940, the DC-3 was carrying an extraordinary 80 percent of America's airline traffic. Eventually Douglas turned out 11,000 of the craft.

Like the DC-1 and DC-2, the DC-3 incorporated many design and structural ideas proposed and tested by Caltech's five aerodynamicists—von Kármán, C. Millikan, A. Raymond, M. Klein, and E. Sechler. Klein had designed a wing joint that the Douglas engineer Lee Atwood adapted for use on the DC-1 and further improved on the DC-2 and DC-3. Raymond credits Sechler with working out much "of the basic structural design parameters on stiff and monocoque metal wings and fuselages . . . in [Caltech's] structures lab." Caltech's engineers recommended cowlings, designed by the National Advisory Committee on Aeronautics (NACA) on the DC-3's engines, and mapped out a better location for the engine nacelles on the wings; and Douglas engineers and designers relied on wind tunnel data for choosing efficient wing flaps and wing arrangements. Contrary to popular belief, the Douglas series trademark—the swept-back wing—did not (according to Raymond) spring from Caltech's wind tunnel tests.

Douglas's all-metal structure (the first American all-metal airplane had been a Ford trimotor in 1926) marked a major step toward safer flying. When asked by an interviewer why the Douglas Aircraft Company had specified the Northrop Alpha design, Maj Klein replied,

> I think what they meant was a metal airplane that was reasonably simple and well thought out, in which the parts were accessible and repairable. Because the only metal airplane they had was the Stout-Ford design, and that was a corrugated thing, and was pretty difficult to repair apparently. The Fokker had wooden wings, and they were just incredibly bad. You see, the older airplanes they had to overhaul every thousand hours. . . . With the sheet metal airplanes, they started

overhauling them every thousand hours and couldn't find anything to do. So they went to two thousand, then to three thousand, and finally up to five thousand.

The hours spent testing models in Caltech's wind tunnel had not been wasted.

There is a deceptively skimpy record of interaction between the Guggenheim Laboratory and local aircraft companies in the 1930s. In order to appreciate the strength of those bonds, we have to turn to an aircraft company located elsewhere—the Boeing Airplane Company, in Seattle, Washington. Boeing was just too far from Pasadena for it to do business by telephone on a weekly basis. Consequently, the quantity of correspondence is much larger than that, for example, with Douglas, Northrop, Consolidated Aircraft, or Lockheed, all located in southern California.

In 1931, Boeing and Caltech began to write back and forth about the testing of the company's planes and airfoils in GALCIT's wind tunnel. While Caltech guaranteed that the results of its tests would remain secret for one year, there was also some leeway as to what should be kept "secret." Maj Klein, for example, informed Boeing officials that Douglas was using highly polished metal airfoils to obtain higher lift coefficients. On another occasion, Klein shared with Boeing the news that GALCIT workers had found the Northrop Alpha low-wing monoplane to have aerodynamic characteristics very nearly as good as those of a high-wing plane. In still another letter, Klein explained how he and his co-workers had improved the Alpha's lift coefficient.

Information about Boeing's test results also leaked out. In response to complaints about this from Boeing in 1932, Klein offered to run the tests requested during the Institute's "school vacation [when] it will be easier than usual for us to maintain secrecy." The problem persisted, however, much to the dismay of Boeing's chief engineer, C. N. Monteith. After Arthur Raymond was promoted to chief engineer at Douglas, in 1934, Monteith and Raymond compared notes. No stranger to "malicious reports" about his firm's spying on Douglas and vice versa, Monteith nevertheless admitted to being a little uneasy about GALCIT's close ties to Douglas in particular. "As far as Cal. Tech. is concerned," he wrote Raymond,

> we are naturally a bit suspicious because, being practically in your own backyard, it is to be expected that the loyalty of the staff of Cal. Tech. is largely yours. As far as allowing data to get out is concerned, we have to take their word for the fact that it does not, but the fact remains that when we had that series of airfoils tested, we received fragmentary data on the results through an individual at Wright Field before we received the data from Cal. Tech. What the explanation is, I do not know.

Nor is it likely he ever found out. But when Monteith ordered a new round of airplane tests in Pasadena later that year, he bluntly reminded Klein of "the necessity for keeping any information with regard to these tests entirely confidential." By 1941, Boeing was poised to build its own wind tunnel.

In many instances, wind tunnel work was initially undertaken at the request of a sponsor and at the sponsor's expense. But many dynamic problems, such as the influence of running propellers on airplane performance, effects of turbulence on maximum-lift coefficients, and control hinge moments, took on a research life of their own. Tests were subsequently amplified by the Guggenheim Laboratory's staff to a degree not at all contemplated when the work was started. Serving two masters can also lead to divided loyalties. For example, the day before leaving for a scientific meeting abroad, Clark Millikan sent Donald Douglas a copy of the paper he planned to present, which, he reminded Douglas, "I discussed with you some weeks ago, and which contains a good deal of material emanating from your factory. I would suggest that you might turn it over to Ray [Arthur Raymond] or one of his engineers to see if there is any material which, from your point of view, should be altered." On such short notice, it was impossible for Douglas to reply soon enough to stop Millikan.

By 1934, Clark Millikan had risen to the rank of associate professor of aeronautics. Besides teaching and research, his main work consisted of supervising the industrial use of the wind tunnel. Millikan was a staunch defender of the advantages of cooperative activities between an independent research laboratory (such as Caltech's Guggenheim Laboratory) and the practical concerns of airplane makers. As he remarked on one occasion in 1936, "the constant contact of the academic members of the laboratory staff with the immediate problems of practical designers has a very beneficial effect upon their research and academic activities by keeping the latter up-to-date and in the current of actual aeronautical research." Daniel and Harry Guggenheim's faith in the practical side of aeronautics had not been entirely misplaced.

At the laboratory's own weekly academic seminar, Millikan added, it was not unusual to have eight or ten visiting engineers representing four or five factories. Moreover, the visitors actively participated in the discussions. The training of aeronautical engineers kept pace with the laboratory's hectic work schedule. Between 1929 and 1939, Caltech awarded 171 advanced degrees in aeronautics, including 21 Ph.D.'s.

Of course, what Caltech's aeronautical laboratory provided, the aircraft companies didn't have to provide. Well into the 1940s, very few such companies had research groups comparable to those maintained by the electrical industry, for example. Advanced research in aeronautics was done in government laboratories, under the auspices of the National Advisory Committee on Aeronautics, and at some graduate schools, including Caltech.

At the outset, the GALCIT wind tunnel was thought of as a bridge to the industrial world. Half of its time was budgeted for research problems. The other half was expected to be used up in testing new airplane designs for southern California aircraft manufacturers.

The Great Depression and Guggenheim's refusal to provide any endowment funds to cover the laboratory's running expenses altered this sunny picture. The fund's three objectives had been to increase the pool of trained aeronautical engineers, to stimulate basic research in aerodynamics, and to foster cooperation with the aeronautical industry. By the late 1930s, the wind tunnel's usage schedule needed to be revised. Local industry as well as manufacturers from other parts of America and from Europe used it so much "that for the time being," Clark Millikan moaned, "all independent research work has had to be abandoned in spite of a seventeen-hour daily operating schedule." In effect, the busy schedule for the wind tunnel forced students to take a backseat to industrial users.

By 1939, von Kármán was taking a dim view of the situation. At the time, Robert Millikan was trying to get government funding for a second wind tunnel in order, in his words, "to serve the aeronautical industry in southern California." Von Kármán didn't think so well of Robert Millikan's lobbying. "Personally," he wrote a colleague, "I am not very much interested in such a wind tunnel because it would not be possible to make any fundamental research in the tunnel; it would be occupied most of the time by routine testing as is the case now with our present atmospheric wind tunnel."

Moreover, industrial testing tended to emphasize a certain class of problems. "As you are probably aware," von Kármán reminded an airline executive in 1939, "the research and instruction in aeronautics at CIT has in the past been principally directed towards the aerodynamical and structural problems of airplane design. This has been the case largely because of the relatively high concentration in this vicinity of the airplane manufacturing industry." To his way of thinking, the very least the industry could do was support GALCIT's academic programs. In 1940, he put the matter before Donald Douglas.

In his letter, von Kármán noted that Caltech had trained dozens of engineers in accordance with the needs of the local aircraft industry, many of whom had in turn gone on to work for Douglas. Would Douglas be willing now to endow several fellowships for promising graduate students in the field? von Kármán asked. Douglas passed the letter on to Raymond for a response. Raymond was less than enthusiastic about von Kármán's idea. How could we ensure that the recipient of the fellowship would join Douglas and not some other firm? replied Raymond. He went on to say, "It is hard to sell the idea of paying out a sum of money for the education of one or more individuals when others who are obtaining a similar training will later take positions with us without this outlay having been made on them." Raymond's reply, written on the eve of America's entry into World War II, shows that individual aircraft companies were not yet prepared to spend money to benefit the industry as a whole. Douglas and the others might be interested, but only in the event that they personally stood to profit by the arrangement. Nevertheless, Caltech's Daniel Guggenheim Graduate School of Aeronautics played a major role in turning southern California into the aircraft capital of the world.

Summing up the changes he had seen in the airplane business between the end of World War I and the beginning of World War II, Klein once said,

> . . . aeronautical engineering was in the process of a revolution. They were going from the old wooden wings and welded steel tubing and wire and cloth covered wings, to sheet metal. . . . Most aeronautical engineers were designing by analogy, as most engineers do now. With the first sheet metal hull that Douglas built, they had built a truss and then they covered it with sheet metal, and that was nonsense. After the prototype, they decided they'd just leave out all the diagonal members, which they did. . . . But they were thinking in terms of bridge trusses, because the designers were all ex–civil engineers at that time. . . . If you look at the production of airplanes in 1930, you will see biplanes, fixed external landing gear, exposed engines. And in 1940, the same manufacturers will be making cantilevered low-winged monoplanes, with retractable landing gear, with cowled engines, and a completely different, modern-looking airplane.

NINE

♦

Atoms, Molecules, and Linus Pauling

> All my life, I have tried to understand chemical substances and their reactions in fundamental terms, i.e., in terms of their protons and electrons.
>
> —LINUS PAULING, 1946

SCIENTISTS are made, not born; but for many, the decision to pursue science comes early in life. Pauling's calling came in 1914, when he was thirteen. Captivated by the chemical experiments a friend had performed for him one day after school, Pauling built his own chemical laboratory at home, scrounged chemicals and equipment from an abandoned steel company on the outskirts of Portland, found some textbooks, and set about repeating various experiments. Even then, he was bothered by "the fuzziness of chemistry, the vagueness of chemical ideas." More important, he wanted to understand the chemical reactions going on in the flasks and test tubes. "I was simply entranced by chemical phenomena," he later wrote, "by the reactions in which substances disappear and other substances, often with strikingly different properties, appear; and I hoped to learn more and more about this aspect of the world."

The opportunity arose, in 1919, in an unexpected way. Then a sophomore at Oregon State Agricultural College (now Oregon State University), majoring in chemical engineering, Pauling dropped out of school between his junior and senior years for lack of money. For a while, he worked as a milkman, then as a paving-plant inspector. To earn his keep at college, he had chopped wood, mopped floors, and helped out in the chemistry department. That November, Pauling received a call from the school, inviting him back to teach quantitative analysis, a course he had just completed himself the previous spring. He also gave lectures in the evenings for students having problems with freshman chemistry. Following the pedagogy of the day, he taught chemical-bond theory as he had learned it: "atoms were assigned a certain number of hooks and eyes which could be hooked into one another to represent the bonds between the atoms." To the eighteen-year-old Pauling, such a picture of the way atoms combine to form molecules seemed reasonable enough.

Its limitations soon became evident. In the course of the year, Pauling learned about the work of Gilbert Newton Lewis and Irving Langmuir. Lewis, a Berkeley chemist, had proposed in 1916 a theory of the chemical bond based on electrons. Like others before him, Lewis saw the necessity of bridging the gap between J. J. Thomson's discovery that the atom itself had a structure and the chemists' ideas about how molecules are built up out of atoms. To account for the way atoms are held together, by chemical bonds, in molecules, Lewis pictured the chemical bond as a pair of electrons shared between two atoms. Langmuir, working independently at the General Electric Company, extended Lewis's work on the electronic structure of molecules. All this came as a revelation to Pauling, who discovered such matters in the *Journal of the American Chemical Society* and other scientific journals only because his office was also the school's chemistry library. The library had few patrons: mostly the departmental secretary, who kept a typewriter there, and Pauling. From this chance encounter with the chemical literature, the aspiring chemist's quest "to understand the physical and chemical properties of substances in relation to the structure of the atoms and molecules of which they are composed" shaped itself.

In the fall of 1920, Pauling returned to school as a full-time student. There were two chemistry seminars during his senior year—one on the chemistry of fish, given by a staff member, and one on the electronic nature of the chemical bond, by Pauling.

Linus Pauling entered the California Institute of Technology as a graduate student in 1922. In the prior three years, Caltech had taken its present name, raised a substantial endowment, embarked on an ambitious graduate research program, and lured the country's foremost physicist, Robert A. Millikan, out west to become Caltech's first administrative head. For all that, the Pasadena school still had a raw and brassy frontier quality about it. Landscaping of the thirty-acre campus consisted of dried, caked weeds, a run-down orange grove, and a handful of small scrub oaks. There were three buildings (Throop Hall, Gates Chemistry Laboratory, and Bridge Laboratory of Physics), twenty-nine advanced students, including ten in chemistry, and eighteen Ph.D.'s on the faculty. Roscoe Gilkey Dickinson, Pauling's thesis adviser, had earned his Ph.D.—Caltech's very first—in 1920.

Dickinson's speciality at the time was X-ray crystallography, a technique for determining the atomic configurations in crystals. X rays are ultrahigh-frequency light waves that tend to be diffracted by crystals into patterns that, to the expert eye, can reveal the arrangements of the atoms inside, and even the precise distances between them. First observed in 1912, the diffraction of X rays by crystals was to prove an indispensable tool to chemists like Pauling with an interest in structure. But in 1922, Pauling's first year as a graduate student, the field was still short, in his words, on "quantitative information about the interaction of X-rays and crystals." Caltech had jumped into this new research field before any other academic institution in the United States, thanks to Arthur Amos Noyes. The founder and first director of the Gates Chemical Laboratory on campus, Noyes determined the direction of the school's chemical research program, recruited the faculty to carry it out, and handpicked the early crop of graduate students, Pauling included. As the leading physical chemist in America, Noyes truly understood the chemical implications of the new technique. He brought Dickinson to Caltech in 1917 to continue the X-ray work that had already begun there.

With the aid of self-generated methods, Dickinson determined the structure of a number of complex crystals by means of X rays. Knowing of Pauling's interest in crystals, Noyes placed Pauling under Dickinson's tutelage. Then began the search for a suitable crystal for Pauling to work on, no simple exercise:

> After two months work and unsuccessful efforts with fifteen crystals I was rescued by Dickinson, who obtained a specimen of molybdenite from the chemistry stockroom. He cleaved a pebble, . . . glued the cleavage to a glass slide, and cleaved again to obtain a thin plate, essentially undistorted. With this plate he and I . . . determined the structure of the mineral.

From the X-ray data, Pauling and his teacher determined the arrangement of each sulfur and molybdenum atom and the distances between them in the mineral crystal. On simpler crystals, the structural determinations could be made in a matter of months. Once he had mastered the technique, there was no stopping Pauling. He and Dickinson submitted the molybdenite paper to the *Journal of the American Chemical Society* in April 1923; in May, Pauling deciphered magnesium stannide, wrote up his findings, and sent it off to be published. Such abilities were not lost on Noyes, who in the course of a lengthy letter to a longtime friend, reported, "One of our Fellows, who came from Oregon, is also proving quite exceptional." In all, Pauling published, alone or with others, seven papers on crystal structures as a graduate student. In 1925, with the promise of a National Research Council fellowship from Noyes in his pocket, Pauling finished up, submitted five of the papers as a thesis, and received his Ph.D. in chemistry.

But there was more than chemistry to Pauling's graduate

school education. The requirements for a doctorate at the Institute included work in another branch of science or engineering.
As there were many more graduate courses in physics and mathematics than in chemistry—and Pauling really liked these courses
—he had no trouble meeting the requirement. He learned integral
equations, complex numbers, and potential theory from Harry
Bateman, professor of mathematics, theoretical physics, and aeronautics; he learned number theory from the mathematician E.
T. Bell. Richard Chace Tolman, professor of physical chemistry
and mathematical physics, taught him relativity theory, mathematical physics, and statistical mechanics. Quantum theory he
picked up from Paul Ehrenfest and Arnold Sommerfeld, two of
the many distinguished campus visitors from whom Pauling
learned theoretical physics as a graduate student. "The faculty
seems to emphasize physics and thermodynamics and statistical
mechanics and atomic structure rather than chemistry," Pauling
told one of his former teachers in Oregon, adding, "I like the
method, but not everyone does."

Pauling has often said that Caltech made a scientist of him.
It helped that the school was small and private and the caliber of
the professors high. Being young, Caltech could make its own
way in the world of science, write its own philosophy of excellence, choose its own traditions, unfettered by the past. Chemistry
research in Gates Laboratory was seen as complementing the
physics research in Bridge Laboratory, as well as the astrophysics
program at the nearby Mount Wilson Observatory, so it was
perfectly natural that a graduate student in chemistry always went
to the physics seminars and to the meetings of the physics and
astronomy club.

If Pauling didn't know everything, it was a temporary deficiency to be remedied later. Once, when Tolman asked him a
question in class, Pauling replied, "I don't know; I haven't taken
a course in that subject." Nor, in other respects, was he self-
effacing. He and Tolman wrote a theoretical paper together dealing with entropy in crystals. When Tolman asked him whose

name should appear first in the article, Pauling volunteered his own. Pauling later defended his action on the grounds that he had spent much more time than Tolman thinking about the structure of matter in general, and the nature of crystals in particular.

Chemists are traditionally concerned with the properties of substances—their color, smell, taste, and so on—and how these properties change when substances are combined. The search for a means of explaining these things took on new life early in the nineteenth century, when the English chemist John Dalton offered evidence supporting the ancient Greek idea that all matter is composed of atoms. But if matter is made of atoms, what are atoms made of? And what is it in the nature of atoms that gives to substances their colors, smells, and tastes? One way or another, these questions formed the backdrop of the evolution of nineteenth-century physics and chemistry. At the end of the century, the English physicist J. J. Thomson showed that all atoms had a common constituent—it came to be called the electron.

An ingenious experimenter, Thomson redesigned an ordinary piece of glass equipment called a cathode-ray tube. Unlike the other tubes, Thomson's apparatus, which was pumped out and sealed while being heated in an oven, had a good vacuum, the key to Thomson's success. He proceeded to demonstrate that the mysterious rays (popularly known as cathode rays) that streamed from a heated filament in the glass tube could be deflected by electric and magnetic fields. Although the deflection of rays by a magnetic field was old news, Thomson was the first to measure the deflection of cathode rays in an electric field. Thomson's experiment offered compelling evidence that cathode rays were electrically charged. That they were negatively charged followed from the fact that the rays streamed away from the negatively charged filament and toward the positively charged one. When Thomson repeated the experiment, but substituted different gases for the air in the tube before it was baked and evacuated, he got the same results. Varying the metallic filaments did not change the outcome of the experiment either. The combined magnetic and electric fields used by Thomson produced a deflection that depended on the ratio of electric charge to mass in the cathode-ray beam. The finding that that ratio always remained the same led him to conclude that he was dealing with a common constituent present in all atoms, the electron. An understanding of the role played by electrons in the structure of atoms and molecules would unlock answers to the ancient questions of chemistry.

Before the time of Thomson, the chemist and the physicist demanded very different things from an atomic theory of matter. The chemist wanted a theory that would explain how the elements combine and recombine as they are mixed, heated, exposed to sunlight, and so forth. The physicist's atom had to give rise to the kinetic theory of gases, line spectra, and, later, radioactivity. These atoms had little in common, but in the wake of Thomson's work, they merged and became one. For thirty years after his 1896 discovery, physicists proposed, tested, modified, and refined models of the atom on the basis of the known properties of the electron. It soon became evident that the tried-and-true principles of classical physics—Newton's laws of mechanics and James Clerk Maxwell's laws of electricity and magnetism—would not suffice to explain the inner workings of atoms. By the late 1920s, a new and more fundamental set of natural laws had been established— quantum mechanics.

The rules of quantum mechanics govern the arrangement, motion, and energy of electrons. The number of electrons in an atom, in turn, governs its chemical properties. Quantum mechanics explains in a profound way the information summarized in the periodic table of the chemical elements and therefore provides a means of answering the traditional questions of chemistry. Its historical origins notwithstanding, modern chemistry is rooted in physics, and at the core of modern chemistry is the work of Linus Pauling.

In 1925, Pauling was probably better prepared than any other chemist of his generation to apply quantum physics to chemistry. Recalling many years later his attitude in the midtwenties toward chemistry and physics, Pauling wrote, "I felt that the time had come when it would be possible to make a successful attack on many chemical problems by applying quantum theory and quantum mechanics, but I did not feel that I might make a successful attack on some problems in physics itself." His mentor Noyes had some plans of his own for how to succeed.

Since National Research Council fellows were required to forsake their home institution for postgraduate study, Pauling had elected to work with G. N. Lewis at Berkeley. Noyes didn't like the idea, counseling Pauling instead to stay in Pasadena until he had finished writing up his various X-ray researches. After that, Pauling could go to Berkeley, Noyes said. Several months later, though, Noyes seemed to change his mind: Pauling should apply for one of the new Guggenheim fellowships and use it to go to Europe to study. He applied for the Guggenheim, but he began packing for Berkeley. At this point, Noyes rather panicked. Afraid of losing Pauling to the University of California forever, Noyes offered to pay his passage to Europe at once and to support him until he received the Guggenheim. This offer was too good to pass up. Pauling resigned his NRC fellowship, and sailed for Europe in March 1926, passing the next nineteen months in the company of Europe's finest theoretical physicists. While abroad, he was appointed assistant professor of theoretical chemistry at Caltech. Noyes never told his young protégé that G. N. Lewis had come to Pasadena in 1925 to offer him a similar position at Berkeley.

Pauling spent a year in Munich with Arnold Sommerfeld, then a month in Copenhagen with Niels Bohr, and the remaining six months in Zurich with Erwin Schrödinger. Introduced to early quantum theory in Pasadena, Pauling was now at the source for lessons in modern quantum mechanics. In Munich, he attended Sommerfeld's lectures on wave mechanics, Schrödinger's form of quantum mechanics. The question in his mind at the time, he recalls, was "Is quantum mechanics, wave mechanics, . . . sufficiently close to being correct so that if we solve the equation we'll get the right answers in relation to the properties of atoms and molecules?" He didn't have long to wait. In 1927, a quantum mechanical treatment of the molecular hydrogen ion appeared in print. Then, just before he arrived in Zurich, W. Heitler and F. London proposed a quantum mechanical treatment of shared electrons, the covalent bond in the hydrogen molecule. The method greatly appealed to Pauling. He later told an interviewer,

> I thought there was a possibility of doing something better but I didn't know what it was that needed to be done. Here I think I had the feeling that if I worked in this field I probably would find something, make some discovery, and that the probability was high enough to justify my working in the field.

He pressed forward with his own quantum mechanical calculations.

In Zurich, Pauling spent most of his time trying to derive from the new quantum mechanics the interaction of two helium atoms. The solution to the general problem of the nature of the chemical bond depended in a fundamental way, Pauling argued, on the quantitative treatment of the simplest molecules. In Copenhagen, he worked with a Dutch physicist named Samuel Goudsmit on spectral problems, finding the theory useful to his own chemical interests. "Some people," he wrote Noyes while abroad, "seem to think that work such as mine, dealing with the properties of atoms and molecules, should be classed with physics but I, as I have said before, feel that the study of chemical substances remains chemistry even though it reach the state in which it requires the use of considerable mathematics." By the time Pauling returned to Pasadena in the fall of 1927, his quest to formulate a quantitative theory of the chemical bond had begun.

Up to then, the chemical-bond problem had belonged to theoretical physicists. In 1928, Pauling wrote his first paper on the quantum mechanics of the chemical bond; many others followed, including a seven-part series on various aspects of the theory—titled simply "The Nature of the Chemical Bond"—that appeared between 1931 and 1933 in the pages of the *Journal of the American Chemical Society*. When asked in the course of a recent conversation why there was a three-year break between the 1928 paper and the "first significant paper" in 1931, Pauling explained, "Well, there was this gap because I was having so much trouble getting a result that was in simple enough form to be valuable to chemists and to have more significance than numbers that you would get out of a computer nowadays." By 1935, he finally understood—to his own satisfaction—the nature of the chemical bond and its properties.

Building on the rules G. N. Lewis devised for the electron-pair bond, Pauling laid down a powerful new set of rules concerning the chemical bonds of molecules. These rules offered guidance on bond lengths, the angles between bonds, magnetic moments, and other molecular properties. Into this detailed blueprint of the way atoms combine to form simple molecules, Pauling introduced an idea called resonance. According to the principles of quantum mechanics, it is possible to represent molecules as alternating or "resonating" between each of a number of possible states in a wholly unpredictable way. Yet, by allowing the bond to alternate in an unpredictable way, one could predict the properties of the molecule with enormous power and precision. This idea, easy to state but impossible to imagine in the intellectual context of classical, pre-twentieth-century science, formed only one part of Pauling's contribution to understanding the nature of the chemical bond. For his research in this field, Pauling received the 1954 Nobel Prize in chemistry.

In 1932, in the midst of the Great Depression, *Fortune* magazine sent a team of reporters to Caltech. They saw no bread lines or obvious poverty on the campus, now boasting a score of new buildings and an expanded student-to-faculty ratio. What greeted them instead were biologists hunched over microscopes, cosmologists arguing about the expansion of the universe, physicists probing the core of the atom. Dazzled by the nature of the work and the intensity of those engaged in it, the writers went back to New York and wrote lavishly about Millikan's cosmic-ray research, George Ellery Hale's astrophysics program involving the 100-inch telescope, and Thomas Hunt Morgan's fruit flies, among other things. They filed no stories about Pauling's work and by way of explanation noted, "Modern chemistry insensibly merges into physics on the one hand and biology on the other."

Pauling's work in biology had accelerated the merging process. His interest in the field dates back to Morgan's arrival on the campus in the late twenties. A geneticist, Morgan came to Caltech to head the school's new biology division; in his wake, other geneticists followed. Pauling began to audit their seminars, and by 1931 he was invited to give his own. His contributions were welcome. As one of the researchers present recalls,

> Pauling was useful to the geneticists always. In the first place, he could understand what you were telling him that you wanted done, and he could tell you what mathematics to use. . . . An article came out on genetics in German on a mathematical theory of crossing over, which none of us could understand. We asked Linus to give us a seminar on it, so he did, and then went on to give one of his own interpretations.

In the meanwhile, Pauling was a very busy fellow—teaching a course at Caltech on quantum mechanics with chemical applications, running the chemistry division's research conference on crystal and molecular structure, spending a month or so each year as a visiting lecturer in chemistry and physics at Berkeley, and, incidentally, turning down professorships at Harvard and elsewhere. In 1931, he was promoted to full professor; in the same year, he received the American Chemical Society's prize for distinguished research in pure chemistry. He was thirty years old.

In 1932, Pauling submitted a proposal for research in structural chemistry to the Rockefeller Foundation. While the work was planned to deal mainly with inorganic substances, he pointed out the need to determine the structure of organic substances as well. "This knowledge," Pauling noted, as an aside, "may be of great importance to biochemistry, resulting in the determination of the structures of proteins, haemoglobin, and other complicated organic substances." The foundation funded the structural-chemistry project for two years.

In 1934, Pauling learned that Rockefeller did not intend to support molecular-structure research as such any more, but would welcome an application from Pauling for research on substances of biological importance. By then, Pauling's interest had greatly shifted; he had acquired a taste for large molecules, the kind present in living things. One of the first organic substances he worked on was hemoglobin, the protein that colors blood cells red.

It was typical of Pauling's approach to understanding the chemical and physical properties of substances that he saw the problem in structural terms. His line of attack, pursued in collaboration with Charles D. Coryell, a postdoctoral fellow at Caltech, started out with a question about the chemical bonding between the hemoglobin and the oxygen that it picks up in the lungs: "I had initiated the work on magnetic properties of hemoglobin in order to find by experiment whether or not the two unpaired electrons of the oxygen molecule remain unpaired when oxygen combines with hemoglobin." The experiment gave an unexpected answer: there are four iron atoms in the hemoglobin molecule. When the molecule combines with oxygen, each of the iron atoms changes its chemical state, acquiring an extra electron. Clearly, iron is the key to how hemoglobin performs its task.

Fresh from this discovery, Pauling and his co-workers turned to the general problem of the structure of proteins—with notable success. By the early 1950s, Pauling's application and enhancement of chemical techniques hammered out at the start of his career had yielded the first dependable descriptions of the physical characteristics of giant protein molecules.

Why did Pauling's laboratory unlock the secrets of proteins where so many others had failed? To begin with, he approached biology as he had approached chemistry. Before he started working on quantum mechanics, he did not know where it would lead him, although he clearly saw the need to bring atomic and molecular chemistry into line with physics. In the end, quantum mechanics provided a method for calculating the properties of small molecules. Moreover, it provided a basis for making qualitative predictions where quantitative studies had had no success. In the same way, when he started working on the large molecules of biology, he thought of the structure of proteins as a problem in atomic structure. "All my life," he remarked in the course of an interview in the midforties, "I have tried to understand chemical substances and their reactions in fundamental terms, i.e., in terms of their protons and electrons." Pauling's way of seeing the world had carried over into biology.

Proteins are made up of amino acids held together in polypeptide chains by linkages between their carbon and nitrogen atoms. Although it had been known since the turn of the century that proteins are long chains of amino acids, the chain's structural features had stumped biochemists and crystallographers alike. Every field has rules, but, as an outsider, Pauling could—and did—simply disregard the prevailing orthodoxy concerning protein structure investigations. Instead of training the X-ray apparatus on proteins directly, Pauling and his collaborators went to work on crystallizing the amino acids that form the proteins, then determining their crystal structure. Other workers in the field mostly rejected his approach to the problem.

But Pauling the structural chemist felt sure that piecing together the molecular dimensions of these simpler substances would lead ultimately to an understanding of the complex fibrous proteins. He had his own style of thinking: ". . . I try to identify myself with the atoms. I ask what I would do if I were a carbon atom or a sodium atom under these circumstances. . . . From the picture worked out we develop an X-ray pattern." He and his co-workers then checked the structural picture by X-ray diffraction methods. Used first on glycine, the simplest amino acid, and then on more complicated amino acid molecules, the model-building method proved astonishingly successful. It led to the discovery of the role hydrogen bonds play in amino acid crystals and, ultimately, to the assignment of a spiral structure, subsequently called the alpha helix, present in ordinary hair, hemoglobin, fingernail, and many other proteins.

To determine the position of the atoms in the coiled chain, the Caltech group measured the interatomic distances and the bond angles. "We took these distances very literally," Pauling later reported, "and calculated the forces accordingly." From the structural information obtained, they derived two spiral configurations, the gamma helix and the well-known alpha helix. To aid them in studying the possible configurations of the polypeptide chain, the researchers also built precisely machined, three-dimensional molecular models. Before Pauling, crystallographers had assumed that the number of molecules in each turn of a spiral structure had to be integral. But the atomic structures that came out of the Caltech group's calculations demolished this idea. The number of amino acid components in each turn of Pauling's alpha helix is 3.7, not an integer. As one of Pauling's overseas colleagues put it, "He knew his atoms and their various states and binding conditions so well that he was prepared to break with what are after all only conventions . . . if they could not be fitted into these regularities." In large measure, Pauling's successful attack on the structure of proteins owes not a little to his ability to translate the problem into one of molecular structure.

Pauling is one of the founders of modern molecular biology. The discovery of the alpha helix (where intermolecular hydrogen bonds provide structural stability) played a crucial role in the work of Watson and Crick, who discovered the double-helix structure of DNA (where hydrogen bonding between base pairs provides structural stability), at Cambridge University in 1953. While the double-chain idea did not originate with Pauling directly, the University of London X-ray crystallographer J. D. Bernal once remarked, "his helical hypothesis was essential to its formulation because it made it possible to predict such a structure and to predict it accurately." In structural chemistry and in biology, Pauling's work in the interwar years and beyond fell in the borderland between physics, mathematics, and chemistry. The successful application of techniques in these fields to problems of life itself constitutes one measure of his scientific prowess. From the beginning, Pauling's scientific qualities set him apart. In the early thirties, Noyes is reported to have said of his successor at Caltech, "Were all the rest of the Chemistry Dept. wiped away except P., it would still be one of the most important departments of chemistry in the world."

Shortly after World War II, Pauling wrote a short piece on molecular problems in which he said,

> . . . although we can weigh them [protein molecules], we do not yet know the exact architecture of a single one of these protein molecules. This then is the great problem of modern chemistry— the determination of the molecular architecture of the proteins and other complex constituents of the living organism. *This problem must be solved.*

The problem would be solved, and in no small measure the solution was based on the foundation laid down by Pauling. In a century of increasing specialization in science, Linus Pauling bridged the gap between quantum mechanics and life itself.

TEN

The Thomas Hunt Morgan
Era in Biology

The establishment of a Department of Biology, rather than the traditional departments of Botany and Zoology, calls for a word of explanation. It is with a desire to lay emphasis on the fundamental principles underlying the life processes in animals and plants that an effort will be made to bring together, in a single group, men whose common interests are in the discovery of the unity of the phenomena of living organisms rather than in the investigation of their manifold diversities.

—THOMAS HUNT
MORGAN, 1927

THE BIOCHEMIST HENRY BORSOOK liked to tell about a conversation Thomas Hunt Morgan had with the physicist Albert Einstein, a campus visitor in the early thirties. At a point in the conversation, Einstein supposedly asked, "What in hell are you doing in a place like this?" "The future of biology rests in the application of the methods and ideas of physics, chemistry, and mathematics," replied Morgan. The physicist persisted. "Do you think you will ever be able to explain in terms of chemistry or physics so important a biological phenomenon as first love?" "What did you say to that one?" Borsook asked Morgan afterward. "I tried to explain something about the connection between sense organs and the brain and hormones." "You didn't believe that yourself, did you?" Borsook asked. "No," said Morgan, "but I had to say something to him."

What he "had to say" says a lot about his plans, however. Thomas Hunt Morgan was sixty-two when he came to Caltech in 1928. By then, he had earned a worldwide reputation as a remarkable teacher, a clear writer, and an impressive researcher. In 1933, he would win the Nobel Prize for his discovery of the chromosomal mechanism by which character traits are passed on from parent to offspring through the interaction of genes.

All of that work had been accomplished in one room at Columbia University that held a bunch of bananas hanging in the corner and eight desks crammed into a space measuring sixteen feet by twenty-three feet. In the fly room, as it was known, Morgan had elevated the lowly fruit fly, the *Drosophila melanogaster*, into the most famous experimental organism in the world.

Why did Caltech officials pursue a biologist so near retirement to establish the school's division of biology? The answer to the question begins with the ways and means of Morgan's *Drosophila* group at Columbia.

According to the Russian geneticist Theodosius Dobzhansky, Morgan ran the fly room by his own rules. Traveling on a Rockefeller-financed International Education Board fellowship, Dobzhansky in 1927 arrived in New York from Leningrad thinking that Morgan was "just next to God" and his laboratory "close to Heaven." To his dismay, he found what he called "a very small, poorly equipped, and positively filthy" laboratory, run by a man obsessed by "pathological stinginess." The Morgan operation made Dobzhansky's laboratory facilities back home in Russia look very good by comparison. Morgan's longtime co-workers Calvin

B. Bridges and Alfred H. Sturtevant sat and worked in the same room, along with graduate students, postdoctoral fellows, like Dobzhansky, and assorted visitors, ranging from Yoshitaka Imai, a Japanese geneticist to Alexander Weinstein, a recent fly room Ph.D. Often, all the desks were occupied, including the two reserved for guests.

Cleanliness was unknown here. The workers competed for space with cockroaches that reproduced in awesome numbers. Contributing to the squalor was Morgan's habit of squashing his flies (after he'd finished counting them) on his counting plate, which he left unwashed on his desk. As the pile of lifeless flies grew, so did the mold it attracted. Along one wall stood a kitchen table used by the student hired to wash bottles and prepare fly food. This area of the room was Morgan's only concession to hygiene.

Bridges, an unremitting tinkerer, sat at a desk covered with odd-looking pieces of apparatus he had made from items at hand. Capable of long periods of routine work and intense "fits and spurts" of ingenuity, Bridges gradually overhauled Morgan's primitive laboratory techniques: he designed a binocular microscope (most workers used a hand lens to examine flies; Morgan used a jeweler's loupe), invented new ways to etherize flies, developed new incubators, improved culture bottles, and whipped up alternative foods for flies.

Bridges had his faults, but jealousy, according to Morgan, was not one of them: "In fact, one of his most admirable traits was his freedom from priority claims of any kind." Morgan first met Bridges in 1909 when he took Morgan's courses in general biology and embryology. Hoping to find research work, Bridges put himself through Columbia on scholarships and odd jobs. Morgan hired him in 1910 to wash glassware, but gave him a desk and promoted him to the job of breeding flies and looking for mutants after Bridges spotted a fly with bright eyes in a dirty bottle. Bridges excelled at finding new mutants, which he "immediately announced." This skill (Sturtevant insisted he "had the best 'eye' for new types") paid off in 1916, when he published, in the first issue of *Genetics*, a detailed paper dealing with flies that had extra and missing chromosomes. Not only could Bridges explain these exceptions; he provided convincing proof of the chromosome theory of heredity. Bridges delighted in building up and studying the *Drosophila* stocks and mapping the position of mutant genes in each chromosome. He "was so good at this that he contributed many more mutants than did the rest of us," Sturtevant once admitted.

Sturtevant owed his desk in Morgan's laboratory to a childhood passion for recording the pedigrees of horses. A book on Mendelism that he read as a college sophomore opened new worlds, he later wrote, "for I could see that the principles could be applied to the inheritance of colors in the horses whose pedigrees I knew so well." He wrote up an account of his findings and submitted it to Morgan, his biology teacher. Much impressed, Morgan urged the young man to publish the account, which he did in due course. Sturtevant always believed this was the reason why in the fall of 1910 Morgan invited him into the laboratory and gave him a desk and some *Drosophila* to work on. By that time, Sturtevant knew for sure that he wanted to do genetics.

Sturtevant was the bookish one. Piles of books and reprints, stacked high, covered his desk. In the course of cleaning the room one summer, so the story goes, a workman found it necessary to rearrange some of Sturtevant's papers, uncovering a shriveled mouse.

It didn't take Dobzhansky long to discover what made the fly room tick. He later told an interviewer, "So this one room had six people working in it, a situation which doubtless had a great many advantages, particularly for a foreign guest. You can ask anyone a question you wish to enlighten yourself on any problem which arises. You also listen to the conversation between the people. As far as training is concerned, nothing better can really be imagined." Jammed with people and paraphernalia, Morgan's laboratory, in short, was an ideal training ground for budding experimental biologists, good for everything from selecting projects on which to base Ph.D. research to testing new techniques and analyzing experimental data.

Sturtevant tells a similar story. "Everybody did his own experiments with little or no supervision," he wrote on one occasion, "but each new result was freely discussed by the group." Morgan's *Drosophila* group did not go in for organized coffee breaks, nor did it set aside a certain time of the day for laboratory discussion. "Instead, recalled Sturtevant, "we discussed, planned, and argued—all day every day." He added, "I've sometimes wondered how any work got done, with the amount of talk that went on." But Morgan did have one cardinal rule: you had to pick your own research topic.

To do otherwise could be academically fatal, as one aspiring *Drosophila* geneticist, Edgar Altenburg, discovered. Having been given desk space in the fly room, Altenburg asked Morgan to suggest a fruit fly problem for graduate work. Close by Morgan's office was the aquarium room. He took Altenburg there, dipped his finger in a tank of stagnant water, and held it up to the light. "There are a lot of Daphnia in here," he said to Altenburg. "Why don't you work on them?" Humiliated by the experience, Altenburg quickly switched to plant genetics.

Dobzhansky nearly made the same mistake. Once unpacked, he wasted no time in asking Morgan "to suggest a topic." "After all I was coming from afar, and although I knew what they had published earlier, I didn't know what they were doing at the time, less still . . . what they were planning to do" in the future, he said, adding, "I did not know how foolish that was." At first, Morgan brushed him off with a joke. When Dobzhansky asked him again for something to read, "the Boss" reached into his desk, took out a reprint dealing with the effects of temperature on the development of *Drosophila* (a subject of no interest to the Russian biologist), handed it over to the newcomer, and turned away. A zoologist by training and a geneticist with an expert's knowledge of the natural variations in lady beetles and with a keen desire to study problems of evolution, Dobzhansky quickly made his peace with Morgan's managerial style. Morgan "thought that everybody should work on the problem which he sees fit," and Dobzhansky knew he was "perfectly capable of choosing" what he wanted to do in Morgan's laboratory. It was a perfect fit.

Morgan was a southerner, born in Lexington, Kentucky, in 1866. His maternal great-grandfather was Francis Scott Key, author of "The Star-Spangled Banner," and his father's brother, John Hunt Morgan, had been a notorious Confederate general. Preferring natural history to politics, Thomas Hunt Morgan in his youth combed the backwoods and byways of rural Kentucky and western Maryland, collecting fauna and fossils. One summer, he earned his keep working for the U.S. Geological Survey, tracing coal seams. After graduating in 1886 with a B.S. in zoology from the University of Kentucky, Morgan spent the summer months working in a marine biological laboratory at Annisquam, Massachusetts, and then enrolled as a graduate student at Johns Hopkins, earning his Ph.D.—his dissertation involved studying different species of sea spiders—in 1890. By the time Morgan joined the Columbia University faculty, in 1904, he was known far and wide for his work in experimental embryology and regeneration.

But the studies that brought lasting fame to Morgan were those connected with *Drosophila*. He had begun breeding fruit flies in 1908 in an effort to determine what role—if any—chromosomes played in the transmission of physical traits from one generation to the next.

Morgan began his research with *Drosophila* just as biologists were beginning to appreciate for the first time the long-neglected findings of the nineteenth-century Austrian monk Gregor Mendel. Mendel's genetic experiments on plant hybrids, published as a short report in 1866, led him to conclude that traits in garden peas such as seed shape, pod color, and plant stem length were determined by fundamental units of inheritance, which he called "elements." Alternative forms existed for every hereditary trait as well: round seeds and wrinkled; green pods and yellow; short stems and long—one of which always stood out decisively in the pea plant. Today every schoolchild knows how Mendel, toiling alone in his monastery's vegetable patch, theorized the existence of dominant and recessive traits, through repeated crossings of innumerable pea plants. Yet Mendel's work was ignored and forgotten until 1900, when it was rediscovered independently by three botanists.

Nevertheless, many researchers, Morgan included, were initially reluctant to accept the notion that Mendel's "elements" (the term *gene* was coined only in 1909) were parts of chromosomes, unless they had evidence rooted not in statistical studies of monastery peas but rather in observable laboratory phenomena. Thus, the question facing Morgan and like-minded colleagues was two-fold: First, how far could Mendel's work be taken as an authentic description of heredity in organisms? Second, how correct was the theory that chromosomes were indeed the physical basis of inheritance? The validity of the chromosome theory took Calvin Bridges—first Morgan's student and later his collaborator—only a few years (1914–16) to establish. The evidence needed to convince Morgan that Mendel's "genes" were indeed carried on chromosomes took longer to accumulate.

Of this period (1910–11) in Morgan's scientific life, the biologist John Moore has remarked, "Whereas it was difficult for Morgan to accept the data of others in suggesting that genes are parts of chromosomes, it was not nearly so difficult when his own data showed the same thing." In the end, Morgan's studies with *Drosophila* convinced him of the necessity of associating specific hereditary characteristics with specific chromosomes. He equated Mendel's elements of heredity with invisible genes at known locations in visible chromosomes and in the process created a new science of genetics.

Fruit flies have been called the geneticist's best friend. They reach maturity quickly, reproduce themselves frequently, and are inexpensive to rear. Shortly after he had begun research on *Drosophila*, in autumn of 1910, Morgan recruited Bridges and Sturtevant, both then still undergraduates, to help with the fly work. Two years later, he brought a graduate student, the physiologist Hermann J. Muller, into the fold, an association that was to have less happy consequences. Muller was a good scientific choice—his work was clearly outstanding—but he and Morgan made a poor match in temperament. From the first, Muller's relations with the group, and with Morgan in particular, were strained. Part of the problem was the pecking order. Scientific decorum mattered a great deal to Muller, who came to feel that others in the fly room got credit for *his* ideas and experimental work. In conversations with the psychologist Anne Roe in the 1940s, Sturtevant testified that "Muller was a very essential part of that group," adding, "We didn't see eye to eye but I got a lot out of him." By the time Morgan moved his laboratory, *Drosophila* stocks, and research group to Pasadena, in 1928, Muller had long since left Columbia and launched his own research team at the University of Texas in Austin, where he became the first geneticist to demonstrate that X rays cause mutations. Like Morgan, Muller eventually won the Nobel Prize, but the personal rift between the two was never entirely repaired.

The discovery that genes are arranged in a single line in chromosomes like beads strung together on a loose necklace was made by Sturtevant in 1911. At the time, he and Morgan had been talking about the meaning behind some diagrams by H. E. Castle of coat colors in rabbits. The diagrams, they decided, were meant to be a representation of the spatial relationships of the genes on a given chromosome. How nice it would be to figure out the geometrical relationship between genes and chromosomes! "I think I can do it," Sturtevant told Morgan all of a sudden. "I suddenly realized," he recalled fifty years later,

> that the variations in strength of linkage, already attributed by Morgan to differences in the spatial separation of the genes, offered the

possibility of determining sequences in the linear dimension of a chromosome. I went home and spent most of the night (to the neglect of my undergraduate homework) in producing the first chromosome map, which included the sex-linked genes y, w, m, and r, in the order and approximately the relative spacing that they still appear on the standard maps.

He published his results in 1913. Sturtevant later told an interviewer that the discovery of the linear arrangement was the most exciting thing he had ever done scientifically. He went on to make other significant discoveries both at Columbia and at Caltech, but none came close to matching the thrill of his "first job on *Drosophila.*"

Two years later, Morgan, Sturtevant, Muller, and Bridges joined forces to produce the first textbook on *Drosophila* genetics, which they entitled *The Mechanism of Mendelian Heredity.* A landmark in the history of twentieth-century biology, the volume quickly became the bible of the new science of genetics. In the hands of Morgan and his co-workers, the genetics of *Drosophila* involved a rigorous and experimental search for the secrets of life that lay sprinkled within the chromosomes of a tiny fly.

By this time, Morgan had largely turned his attention elsewhere. He left the day-to-day operations of the fly room in the care of his students, who were technical virtuosi of the first order. A great synthesizer, Morgan distilled their findings, popularized their work, and shouldered the responsibility for bringing their results to a wide, frequently nonscientific audience. In fact, by the time he left Columbia in 1928, Morgan could no longer follow in detail what his younger colleagues, Bridges and Sturtevant in particular, were doing.

This happens to scientists, even to the best of them. And it's often sad to observe. In Morgan's case, however, distance worked to his advantage. As one geneticist familiar with Morgan's working habits put it, "For Morgan himself the *Drosophila* work was only one aspect of a biologist's searching." Professor of experimental zoology at Columbia for twenty-four years, Morgan in 1928 was still searching, as is plain from the following lines he wrote to George Ellery Hale, shortly after accepting the Caltech job:

> . . . I am writing to you something of the ideas that are shaping themselves in my mind about the organization of our biological work. Would it not be a good plan to think in terms of "The Biological Laboratories," rather than of a "Biological Department." This would allow greater freedom in giving each group an independent footing and allow greater flexibility in the future. As I have intimated to you, I think, I have no ambition to "boss the job," but rather to get together the best men available, to settle down to my own work, and then do all I can to coordinate and help matters forward along constructive lines.
>
> Our program, when we get it going, should speak for itself. . . . And, while I am anxious to emphasize the dynamic or physiological character of the work, I shall try to avoid the criticism that we are leaving the older and less important sides of biology in the background. This can best be done, perhaps, if we point out that we are not so much attempting to duplicate work that is being done well elsewhere, as in furnishing opportunities for the more advanced and less well developed lines of modern research.

Morgan's days of "the fly room" mentality were behind him. "Only through an exact knowledge of the chemical and physical changes taking place in development can we hope to raise the study of development to an exact science," Morgan told Caltech's elders in 1927, shortly after they had approached him about organizing work in biology in Pasadena. "The best chance," for success, he indicated, would be "to put some physicists in the biological laboratory, and some biologists in the physical laboratory."

Morgan's prophetic remarks set the tone of biology at Caltech for the next half century. It was, for example, Caltech's physicist turned biologist, Max Delbrück, himself winner of the Nobel Prize in 1969, who helped lay the foundations of modern molecular biology, and the brave new world that we've only begun to glimpse.

Morgan kept his word to Hale "to get together the best men available," starting with three he knew—Sturtevant, Bridges, and Dobzhansky. He also recruited as teaching fellows three graduate students from Columbia, including Albert Titlebaum. From the University of Michigan, he plucked Ernest Anderson and Sterling Emerson, both Ph.D.'s in plant genetics, but well versed in the genetics of animals as well. Anderson, thirty-seven, came as an associate professor of genetics; Emerson, twenty-nine, as an assistant professor. Another geneticist from Columbia on Morgan's short list, Alexander Weinstein, did not get to Caltech—in a way that says much about the school's early ways.

Weinstein, Morgan told Millikan in the spring of 1928, had been working in the fly room and had just successfully repeated Muller's use of X rays to induce mutations. If appointed to Caltech, he would continue the work in Pasadena and teach the introductory course in biology. Morgan was proposing to make Weinstein an assistant, perhaps even an associate, professor, at an annual starting salary of $3,500. Emerson's starting salary was $3,800 and Anderson's $4,000. As professor of genetics, Sturtevant received $6,500, placing him near the top end of the Caltech pay scale; Morgan was hoping to raise Sturtevant's salary to $7,500. (As head of the new department of biology, Morgan himself made $10,000, the same as Millikan and Noyes.)

A scientist "with distinct literary ability," broadly trained in biology, fluent in mathematics and physical chemistry, Weinstein ("a fine type, not aggressive") struck Morgan as the right man for the Caltech job. "I have hesitated a long time before bringing his name forward," Morgan admitted in his letter to the school's head, "but I think for the position proposed he is the most suitable man at present available."

No one on the faculty, save for members of the National

Academy of Sciences, replied Millikan, made more than $7,000 a year. In Sturtevant's case, Millikan agreed, Morgan might have to pay that much or more to get him, but he might first try less expensive inducements—paying traveling costs to meetings back east or moving expenses. Weinstein's case was scarcely different. Millikan had at least "three brilliant young men" in the assistant professor ranks, all making between $3,000 and $3,300. Offer him $3,500, Millikan counseled, but "then give him a chance to match his pace to an associate professorship with these other men of about his age." In any case, he left all decisions about rank and salary in Morgan's hands.

Morgan did not change Sturtevant's starting salary. He offered Weinstein $3,500 as an assistant professor, which the seasoned fly room veteran refused, pointing out that Emerson, barely out of graduate school, was making more and that Anderson and Sturtevant, who were about his age, were getting higher faculty positions. Morgan bristled. "I . . . consider the matter finished, as I do not think we want to have a man who makes points like that," he informed Millikan by letter that May. Too cheeky perhaps for Morgan's taste, Weinstein went on to teach genetics at Minnesota, then branched out into zoology and the history of science at Johns Hopkins. He later taught physics at City College of New York and eventually wound up at Harvard.

Another possible appointment in biology was Leonor Michaelis, a prominent biophysicist at Johns Hopkins. Michaelis, however, had several strikes against him, according to Caltech's new biology head. One was his apparently pronounced ethnicity. In the same letter of 1928 to Millikan recounting his dealings with Weinstein, Morgan lamented that Michaelis already had "collected about himself a few young Jews." "He himself is markedly Semitic," added Morgan. "I have my doubts whether we should want to start under these conditions, and shall make no moves." Morgan recommended against hiring Michaelis. Fifty years later, Leonor Michaelis's daughter read the discussion about her father in the Caltech archives and said "it was shocking" to learn that the call never came, because, as she put it, "he was a Jew." People, scientists included, rarely take the time to write shocking letters any more; they simply talk on the telephone instead.

Dobzhansky, who worked side by side with Morgan for many years, described him as a biologist with a razor-sharp mind, "a man of wide education which should have made him very broad, but curiously enough, did not." "In many ways," his Caltech colleague once recalled, "he was a very contradictory person." He'd had a number of Jewish co-workers at Columbia—Weinstein, Muller (he claimed one Jewish ancestor), Tyler (born Titlebaum, he changed it after moving to Pasadena)—and he brought many others to Caltech, including Henry Borsook, Jack Schultz, and Norman Horowitz, who many years later said,

> The question of Morgan's alleged anti-Semitism bothers me. I was closer to him than most graduate students during 1936–39, because he and Tyler and I spent every weekend at the marine station. I never noticed any anti-Semitism whatsoever on his part. On the contrary, he was always nice to me, and I have always believed that it was he who got me a National Research Council Fellowship when I finished my Ph.D. in 1939.

"But time and again he would make, especially when irritated, anti-Semitic remarks of the most crude sort," remembered Dobzhansky.

Morgan had a reputation for making outrageous remarks, for teasing those with different beliefs. "But he was never mean," insisted Sturtevant. Robert Millikan was often the butt of Morgan's gibes, for, unlike the atheist Morgan, Millikan was a pious Protestant. But Morgan's penchant for saying the wrong thing eventually caused an international flap in scientific circles. In 1934, Morgan went abroad, partly to pick up his Nobel Prize in Stockholm, partly to recruit new staff members. As Morgan had told a Rockefeller Foundation official beforehand, he wanted "to look over the ground at first hand and make sure that the men we have in mind are the kind we are looking for." While in London,

Morgan attended an elegant reception hosted by the Royal Society. "He has announced to all who will listen," an eyewitness later reported, "that the Rockefeller Foundation has given him money to secure the services of a physiologist. He is combing England and the Scandinavian countries to find one who is not Jewish, if possible." The informant added, "From the English reception of this announcement, I am inclined to believe that he will have difficulty in finding a first-rate Englishman who will be willing to go to Pasadena." Indeed, Morgan was unable to recruit anyone in England or Scotland. Just before sailing home, he hired a Dutchman, C. A. G. Wiersma, who evidently had the right pedigree. According to a Rockefeller official in Amsterdam, Morgan "gave somewhat the impression of being 'desperate.' "

In fall of 1928, the founding members of Caltech's biology division assembled for the first time in Pasadena. Traveling by the Santa Fe Railway, Morgan and his wife, Lilian, also a biologist, left for Pasadena on September 6, 1928, after spending the summer, as always, at the Marine Biological Laboratory in Woods Hole, Massachusetts. Dobzhansky, his wife, and Miss Wallace, Morgan's illustrator and secretary, left Woods Hole and joined him in Pasadena several weeks later, carrying among them "the sacred flame," the Drosophila stocks. Morgan met them at the train station. Bridges turned up soon afterward. Emerson had gone fishing with his father-in-law in Canada meanwhile and arrived at the end of October, a month after the academic term had begun. Sturtevant's wife was expecting a baby, and Sturtevant "thought she had a right to be born in the East." They arrived two months late. Anderson was completing a research trip to Berlin and arrived even later, close to Christmas.

Finding the new biology building unfinished, Morgan and his group set up a makeshift laboratory in the chemistry building, in Arthur Noyes's office. By the time classes began, two rooms in Caltech's new Kerckhoff Laboratories of the Biological Sciences were ready for occupancy. Tucked away by itself in the northwest corner of the campus, the building was a brisk five-minute walk to the physics, chemistry, and engineering laboratories across the quadrangle. A student who took courses there in 1930 recalls that the building "was connected by a boardwalk to the rest of the campus. In the winter, the territory between Gates and Kerckhoff became a sea of mud, known generally on the campus as Lake Kerckhoff."

Caltech officials had promised the biologists that they would not have to teach any courses that first year. But under pressure from the undergraduates, they offered one—in beginning biology—in the spring term. Morgan and Sturtevant divided up the lectures, while Anderson, Emerson, and Sturtevant ran the laboratory associated with it. Partial to Darwin, Morgan in his homework assignments often asked students to read a portion of his masterpiece, *The Origin of Species*, and to write a report on it.

In the genetics laboratory, the atmosphere of the original fly room was soon re-created. A long bench stood in front of the two windows. Dobzhansky and Sturtevant sat at opposite ends of it looking at their flies—Dobzhansky on the left, Sturtevant on the right. "The students sat in between and listened to the wise conversation and contributed to it when they could," one former student remembers.

Intrigued by Muller's X-ray work, Dobzhansky had used his time at Woods Hole during the summer to irradiate flies. He spent fall and winter in Kerckhoff studying the chromosomal aberrations caused by the X rays and arranging them on a chart, using genes as markers. "Just what I expected to see in chromosomes, I don't remember," Dobzhansky later told an interviewer, but he decided one day to look under a microscope at the rearrangements he had projected would be observed between the fly's third and fourth chromosomes. Practiced in dissecting beetles, Dobzhansky removed the ovaries of a young female fly, embedded them in paraffin, and sectioned and stained them. It was a long, tedious process. He looked through the eyepiece.

"Suddenly I saw an incredible thing," he later recalled, "namely, I saw a chromosomal plate which had just one little dot . . . and a chromosome never seen before, a long rod, which clearly meant that a piece of the third chromosome had become attached to the tiny fourth." The ancient dream of the geneticists—direct evidence of the serial order of genes on chromosomes—stared Dobzhansky in the face. "I don't remember whether I emitted a loud yell," he later said.

By spring 1929, Dobzhansky had produced the first cytological map of the fly's long, rodlike third chromosome. To his joy, when he compared the linkage map, which summarized statistical data based on many genetic experiments, to the cytological map, he discovered that the two maps agreed with each other. The ability to predict the inheritance of certain characteristics, Morgan once said, justified the construction of genetic-linkage maps, "even if there were no other facts concerning the location of the genes." Dobzhansky's work in Kerckhoff Laboratory offered irrefutable, direct evidence of the correctness of Morgan's classical theory.

Meanwhile, time was running out on his postgraduate fellowship. Morgan had succeeded in getting Dobzhansky a six-month extension, which meant he had to wind up his research and return home to the Soviet Union at the end of June 1929. One day, he walked into the genetics laboratory and, according to Dobzhansky, "asked the question, 'Dobzhansky'—or rather, he called me to the end of his days, he could not pronounce this name, which of course I don't blame him, it's a devil of a name— he called me 'Dobershansky'—'Would you like to join our staff as assistant professor?' " It took him no time at all to answer yes.

In educational matters, Morgan did not believe in graduate courses; he believed in reading. But even in Morgan's day, graduate students took courses, whether required or not. Seminars abounded, including the general biology seminar each Tuesday night, which Morgan always attended (and at which he introduced the speaker).

In one of these seminars, in 1933, two German researchers

reviewed a paper on the salivary gland chromosomes of flies known as *Bibio*, or March flies, so named because they are commonly abundant in spring. The Germans had observed in the cell nuclei of the larval salivary glands rope-like structures, which they correctly interpreted as "giant chromosomes." A number of geneticists were in the audience that day, Morgan and Bridges included, but they did not get excited by the report. Their attitude changed overnight when Theophilus Painter, a geneticist at Texas, drew and published the first map of these chromosomes for *Drosophila* and pointed out how the banding pattern could be used to study the break points of any chromosomal rearrangements.

As Dobzhansky tells the story, Bridges showed up in his laboratory one morning just thereafter and said, "Dobzhansky, show me the salivary glands." Although he probably knew more about *Drosophila* than anyone else, Bridges had no experience in dissecting larval fruit flies. Indeed, he had never even seen the salivary glands. Dobzhansky dissected a larva and showed the results to his visitor. And Bridges jumped into the study of these giant chromosomes with a vengeance. He set about identifying and extending the number of visible bands of these chromosomes. He went on to produce a series of drawings that are still consulted by fly geneticists. "Bridges's map," the *Drosophila* whiz Edward Lewis remarked more than fifty years later, "is still a masterpiece."

Keen competition existed between the Caltech geneticists and Painter's research group at the University of Texas. In 1934, Painter wrote an indignant letter to the editor of the journal *Genetics*. According to Painter, the two groups were "in a sense competing," and the Texas group had already hit two "home runs": Muller's discovery of X-ray induced mutations, and Painter's own work on the salivary glands. But now he was upset that Bridges had reviewed his salivary gland manuscript (Did Morgan have a hand in this? he asked). Worse still, Bridges's salivary gland work was now "making a splurge in the newspapers"—the *New York Times* had sent a reporter to Cold Spring Harbor to cover Bridges's talk on his salivary gland research—while his, Painter's, contributions had "been belittled." Painter's public howl strikes a familiar chord. It is not uncommon for scientists to believe that their work is underappreciated.

Painter's most pressing complaint, however, concerned a Science News Service press release about Bridges's work, which the magazine's editor had sent to Painter. From reading the marked-up copy, Painter could see that Bridges had corrected his estimate of the size of the salivary gland chromosomes and that the new figures now agreed with Painter's—which had not yet been published. He felt that Bridges had taken advantage of the situation. He wrote in his letter, "I do not intend to reflect in any way on Dr. Bridges. On the other hand, the men in . . . [our] laboratory are unwilling to allow competitors in our field to enjoy the privilege of examining our work a year prior to publication when we have no opportunity to see theirs."

"We are not interested in home runs," Morgan replied, after reading a copy of Painter's letter. As far as he, the dean of American biologists, was concerned, the two research groups were "cooperating," not competing. Far from admitting any wrongdoing, Morgan defended his group's honor, but held out a peace offering: articles submitted by Caltech would now be sent to Austin before publication. In a separate note to the editor, D. F. Jones, Morgan blamed Painter's "outburst" on Muller, noting, "[His] attitude has always been antagonistic to us . . . although he has generally managed to keep this under cover and we have consistently ignored it, treating him in the most friendly way, because we regarded his attitude as wrong and inexcusable." It is reported that Painter later mellowed in his view.

Painter's story bears telling because of Muller's experience several years later. At Caltech, Morgan had laid down the rule that getting papers published was an individual's responsibility, not the department's. But as Sterling Emerson once freely admitted, Morgan's group had no trouble getting him to submit their papers to *Science* and the *American Naturalist*, edited by J. McKeen Cattell, a personal friend. When Morgan submitted a colleague's paper, recalled Emerson, "it would come out in the next issue. It might be any time if you sent it in." Being friends of friends paid off, as Calvin Bridges discovered.

In 1936, Bridges was preparing a paper for *Drosophila* on the Bar gene, a spontaneously mutating gene that reduces the size of the fly's eye. As Bridges drew near the end, word reached the Caltech group that Muller was also working on this gene. Donald Poulson, a graduate student in the lab at the time, told an interviewer in 1978 what he remembered: "I don't know whether I should say anything about this, but I think it's current now—Dobzhansky had had a letter from Russia, from one of his friends, which said, 'Muller has solved the Bar story.'"

Morgan took matters in hand. Bridges's paper was submitted on February 21, to *Science*, and was published a week later. The habitually frugal Morgan, it seems, had wired the entire paper to the journal's editor. So much for Morgan's "cooperation" among *Drosophila* groups. Three months later, a special article by Muller on his own cytological analysis of the Bar gene appeared in the same journal. Muller prefaced his technical remarks by calling "the attention of American readers . . . to the fact that essentially the same findings and interpretation" (presented in Bridges's paper) had already appeared in print in the Soviet Union under Muller's name and that of his two Soviet co-workers. In truth, rivalry is the handmaiden of science, and the quest to be first is a motivating force and a powerful stimulus to creative work. Good scientists are nearly always keen competitors.

Closer to home, Morgan made a magnanimous gesture when in October 1933 he received the Nobel Prize in physiology and medicine. From the proceeds of the prize, some $40,000, he deducted his traveling expenses to Stockholm and back, and divided the rest of the cash prize three ways: one-third of the money went to his own children, one-third went to Bridges's children, and one-third to Sturtevant's.

Biology at Caltech in the thirties was special because of its emphasis on genetics, *the* essential science for the future of biology. Caltech had staked out its claim, and that was Morgan's doing. At other first-class universities, including Harvard and Princeton, genetics took a backseat to other branches of biology, such as comparative anatomy, embryology, and physiology. Caltech was unconventional not only in its choice of discipline but also in its methods of discovery. For one thing, Morgan's approach was completely experimental; no courses in descriptive biology existed. According to people in the department then, Morgan "said that as long as he had any say in this matter, there would never be a class in taxonomy or in morphology."

In setting the intellectual tone for the new division, Morgan was guided by his instincts, and by an outlook much broader than even his best pupils could muster. If he had an ideology, it "was that genetics was the root to finding out how life works," as James Bonner, a former student, put it. Sturtevant said it in a slightly different way: "Morgan's objective in biology was the development of mechanistic interpretations. Anything teleological was sure to arouse his antagonism." There was, Morgan maintained, a rational, physicochemical explanation behind all biological phenomena.

The Caltech plant physiologist and biochemist James Bonner remembers that when he was a graduate student in biology, in the early thirties, a fellow graduate student in physics, "Willy Fowler, who will of course deny this," said to him, " 'Biology? How are you ever going to make a science out of that?' " Morgan came to Caltech to answer that question.

ELEVEN

◆

Astronomy and the 200-Inch Telescope

This project is a sort of scientific heroic poem: and like all truly great poetry, it contains elements of both joy and sorrow.

—WARREN WEAVER, 1947

THE FOCUS of scientific research at the Institute under Millikan during the thirties ranged from *Drosophila* genetics and the biochemistry of vitamins in biology to the theory of turbulence and airplane wing design in aeronautics; from cancer therapy with radiation and the radioactivity of the light elements in nuclear physics to soil erosion and the transmission of water from the Colorado River to Los Angeles in engineering; from the application of quantum mechanics to molecular structure in chemistry and to the introduction of the magnitude scale in seismology.

Each of these was impressive, and important, in its own way, but Caltech's determination to build a 200-inch telescope was in a class by itself. The embryonic idea that grew into the world's largest optical telescope began shortly after the 100-inch telescope on Mount Wilson went into operation. In 1919, the astronomer

Francis G. Pease used the new telescope to photograph the moon, and flushed with success, Pease and George Ellery Hale, the director of the Mount Wilson Observatory, began to dream of a still-larger instrument, one that would truly unveil the heavens. Pease even designed a 300-inch reflecting telescope. A paper of Hale's that appeared in the early 1920s also pointed in that direction. In it, he wrote, "Looking ahead and speculating on the possibilities of future instruments, it may be mentioned that comparative tests of the 60-inch and 100-inch promise well for larger apertures."

Few men were as well qualified to look ahead as Hale. A great builder of telescopes and institutions, Hale had been the guiding force behind the building of three great astronomical instruments, the 40-inch refracting telescope at Yerkes Observatory and the 60-inch and 100-inch reflecting telescopes at Mount Wilson. Praised in 1897 for its revolutionary design, the Yerkes Observatory, at Williams Bay, Wisconsin, was, in Hale's words, "a large physical laboratory as well as an astronomical establishment." There, with the aid of F. Ellerman and J. A. Parkhurst, Hale had studied the spectra of low-temperature red stars. Besides continuing his own research on sunspot spectra, building a specially designed telescope for the sun, he also studied the distribution of calcium clouds (Hale named them flocculi) at different levels in the solar atmosphere, using an instrument he had built for the Yerkes telescope. The discovery of dark clouds of hydrogen vapor (the calcium clouds were bright) followed in May 1903. Eight months later, Hale had decided to build a new solar telescope in a better location out west.

In 1904, Hale founded the Mount Wilson Observatory, in Pasadena, with funds provided by the Carnegie Institution. In an effort to improve the transplanted Yerkes solar telescope, eliminating mirror distortion and air turbulence, Hale built in 1908 a 60-foot tower telescope, with a vertical spectrograph, in an underground shaft, for photographing solar phenomena. By then, he had discovered that sunspots are cooler than their surroundings, and he had found vortex structure in dark hydrogen clouds near sunspots. The discovery of vortices suggested to Hale that the double lines in sunspot spectra—photographed with the 60-foot tower telescope—were not produced by superposed absorbing clouds but rather by magnetic fields. In 1908, Hale compared this astronomical observation with the similar doubling of lines (called the Zeeman effect) obtained with large electromagnets in the observatory's physical laboratory, and demonstrated for the first time the existence of strong sunspot magnetic fields. Working later with the 150-foot tower telescope, Hale attempted to measure the general magnetic field of the sun; he also formulated the law of sunspot magnetic polarities and discovered their reversal in successive eleven-year cycles.

Hale was recognized everywhere as a great scientist, but his physical makeup was not as strong as his mind. For decades, he had suffered from throbbing headaches. In 1923, weakened by a series of nervous breakdowns, the fifty-five-year-old Hale resigned as director of the Mount Wilson Observatory ("I had not the remotest intention of undertaking the organization of another institution of this kind," he wrote later in his unpublished autobiographical notes) and built for himself a small solar laboratory in Pasadena. Thinking about scientific problems and talking to other scientists was not good for his "head," he confided to the Carnegie Institution's president, John Merriam, hence the need to divorce himself from the stimulating intellectual and social life of the observatory. Characterizing himself as "a born adventurer, with a roving disposition" that constantly urged him "toward new enterprises," Hale estimated he had "not enjoyed one third normal working capacity during the last fifteen years."

But the success of the 100-inch telescope and the growth of the staff's interest in stellar physics had whetted the irrepressible Hale's appetite for a still-larger reflecting telescope, to work on fainter stars and probe farther into space. When the editor of *Harper's Monthly Magazine* asked Hale, in 1927, to write a popular astronomy piece for its readers, Hale deliberately wrote a provocative article on large telescopes, in the hope of "interesting someone like Yerkes, Hooker, and Carnegie . . . who might wish to provide the means of penetrating farther into space." Hale did not fail his readers. Emphasizing the grandeur and mystery of space, he introduced the subject this way:

> Like buried treasure, the outposts of the universe have beckoned to the adventurous from immemorial times. Princes and potentates, political or industrial, equally with men of science have felt the lure of the uncharted seas of space, and through their provision of instrumental means the sphere of exploration has rapidly widened. If the cost of gathering celestial treasures exceeds that of searching for the buried chests of a Morgan or a Flint, the expectation of rich return is surely greater and the route no less attractive. . . . Starlight is falling on every square mile of the earth's surface, and the best we can do at present is to gather up and concentrate the rays that strike an area 100 inches in diameter.

The light collected in a telescope is proportional to the area of the mirror, which increases as the square of its diameter. A telescope with twice the diameter of the Mount Wilson 100-inch telescope would collect four times as much light, the giant of that time. This disarmingly simple geometrical relation was a key element in Hale's determination to build a 200-inch telescope, or one larger. With a giant gain in light, finer photographic resolution, and therefore the ability to see farther into space, astronomers would be able to attack problems ranging from the structure of the universe to the structure and evolution of our own Milky Way and of stellar systems beyond and to the evolution of stars and the composition of stellar matter.

In February 1928, Hale asked the editor at *Harper's* to send an advance copy of "The Possibilities of Large Telescopes" to Wickliffe Rose at the Rockefeller Foundation. Then Hale wrote to Rose. Timing was everything, for Rose, the Rockefeller General Education Board's president, was due to retire at the end of June. In his letter, Hale asked Rose if the Rockefeller Foundation would finance a study project to determine how large a telescope mirror it would be feasible to cast. Rose invited Hale to come east for a talk. When Hale called on Rose in New York on March 14, they talked for a while before Rose abruptly asked him, "Do you want a 200-inch or a 300-inch?" "A 200-inch telescope," Hale replied.

The following day, Rose and another foundation official left for Pasadena, where they toured the Mount Wilson Observatory, looked through the telescope, and talked to the astronomers, including Walter Adams, Mount Wilson's director. Afterward, Adams passed on to Hale (who had remained in New York on National Research Council business) a vital piece of information: Rose wanted to put the proposed telescope into the hands of a school, not the Carnegie Institution or the National Academy of Sciences, as Hale had initially proposed. Back from his brief trip to Pasadena, Rose met with Hale twice more in early April before making his first official move. Hale, fighting off a rash of physical ailments, including an ulcerated cornea that kept him holed up in his hotel room during much of his stay back east, reported to his wife, Evelina, in Pasadena, "He has decided to ask his board for about six million dollars for a 200-inch telescope, to be given to the California Institute."

All along, Hale had said nothing to Carnegie's John Merriam about his negotiations with Rockefeller regarding a 200-inch telescope. Now that Rose had made up his mind to push it, Hale could no longer keep Merriam in the dark about his plans. Accompanied by the Caltech chemist Arthur Noyes, he broke the news to Merriam on April 12 over dinner at the University Club. "It was a curious task," Hale later remarked, "to tell Merriam . . . that Rose wants to give it, not to the Carnegie Institution but to the California Institute." Merriam apparently did not object that evening. Then Hale sent Rose a letter summarizing the arguments in favor of building a 200-inch telescope. In it, he assured Rose of the "close cooperation" of Caltech and of the Mount Wilson Observatory in the construction and operation of the new telescope, pointed out the need for a director and a governing board, but avoided the question of who would own the telescope. It was a delicate situation, and Hale was caught in the middle. As he told Evelina,

> They [Rockefeller] do not like the scheme of organization of the Carnegie Institution—he [Rose] thinks, for example, that the Mount Wilson Observatory ought to be an independent institution, with its own endowment and a chance to develop, rather than to depend upon the caprice of someone in Washington. . . . Of course everything is very confidential, especially Rose's views of the Carnegie Institution. As he says, he is speaking only for himself, and he retires on June 30.

Hale left for Pasadena on April 19, hoping the telescope project would not "all blow up."

But it did. In fact, it blew sky high, two days after Hale reached home. Rose had gone to Washington and talked with Merriam on April 25 about the project, only to find him dead set against the whole idea. Rose promptly informed Caltech officials he was going to "drop the matter entirely till all concerned were in agreement."

Carnegie's president was hardly in an agreeable mood. Rose's suggestion that Caltech would make better use of the astronomical instrument if it belonged to them had infuriated Merriam. "As I understand Dr. Rose," Merriam subsequently wrote to the Carnegie trustee Henry Pritchett, "he believes that Mt. Wilson Observatory is eminently fitted to do the work of planning, constructing and operating a two-hundred-inch telescope, but he believes the organization and policy of the Institution inadequate to give proper guarantee for the utilization of such an instrument or the organization of its work in the future. . . ."

Merriam's criticism of the 200-inch telescope project turned into a litany of complaints about Caltech itself. What had the school ever done in the way of astrophysics? he asked Pritchett, pointing out that Caltech had neither course work leading to a

Ph.D. in astronomy nor staff to support an astronomy research program. Merriam was sure that Hale, Millikan, and Noyes had never contemplated starting up work in astronomy at Caltech, much less acquiring a giant telescope. Under the circumstances, Caltech had no business getting involved in such a project.

At least one Rockefeller official, Max Mason, later said that Merriam's hostility to the 200-inch telescope stemmed from the fact that Hale had not offered it to the Carnegie. According to Mason, Merriam would have dismissed any project that "had not been started by himself." President of the Rockefeller Foundation after Wickliffe Rose retired, Mason also volunteered the information that Merriam "disliked Hale." As Mason explained it to one of his own colleagues, "M. instinctively disliked any person to whom he was indebted."

And Merriam was indebted to Hale, for it was Hale who had tapped Merriam in 1920 for the post of Carnegie Institution president. A paleontologist with broad geological interests, Merriam had grown up in Iowa, graduated from Lenox College in 1885, studied geology and botany for a while at the University of California at Berkeley, and then enrolled at the University of Munich to earn a Ph.D. in vertebrate paleontology in 1893. The following year, he returned to Berkeley, and eighteen years later, in 1912, he became chairman of the new department of paleontology. A pioneer in the study of vertebrate paleontology up and down the Pacific Coast, Merriam had also championed the cause of conservation, helped found the Save-the-Redwoods League and served as its president for many years, and taken an active interest in the development of the country's national parks. For all that, Merriam was not an easy man to get to know. "He was by nature not genial, but rather grave and distant," a co-worker later recalled. "His conversation was usually in a serious vein."

In dealing with him, Hale certainly found Merriam less than genial. Having spent a large fraction of his life wheeling and dealing in telescopes, Hale bristled at the notion that Merriam of all people would try to stop him from building the largest telescope in the world. Hale solicited glowing letters of support, not only from Mount Wilson's director, Walter Adams, but also from the world-renowned physicists Albert Michelson and Michael Pupin, which he circulated among the trustees of the Carnegie Institution. Writing as the observatory director, Adams now expressed surprise that anyone would question giving the telescope to Caltech. "I think it is peculiarly appropriate," he wrote, ". . . because of the great strength of its department of physics. The future of the work of such an instrument lies quite as much in the hands of the physicist as of the astronomer." Adams then pointed to the work of the Caltech physicist Ira Bowen. With the aid of a laboratory spectrograph, Bowen had been the first to identify "the unknown lines in the spectra of gaseous nebulae," a feat that ranked as "one of the most brilliant astronomical discoveries of recent years," in Adams's judgment. Bowen's success depended on his knowledge of the rapidly burgeoning field of atomic physics; the nebular lines proved to be oxygen at very low density.

Before packing his bags on April 29, 1928, and returning to the East Coast, Hale had individually approached Pritchett, J. J. Carty, Elihu Root, and other Carnegie trustees. By the time the Twentieth Century Limited had pulled into the station in New York, he had mapped out a battle plan. "I will come back the moment I can safely leave," he promised Evelina. "But I must leave no loopholes this time." By May 8, Hale had each and every Carnegie trustee, including the trustees' president, Root, totally on his side. Pritchett and Root prepared a letter for Rockefeller promising complete cooperation on the 200-inch project, and Pritchett took it to Washington for Merriam's signature. "If Merriam does not want to sign Mr. Root said he wanted him to come here [to New York] at once to see him," Hale reported to his wife. Merriam refused to sign it.

But the anticipated showdown between Merriam and Root on May 10 in New York, which Hale was dreading, never happened. Waiting for Merriam at Pennsylvania Station in New York that day was a delegation of Carnegie trustees consisting of Carty, Elihu Root, Jr., and Frederick Keppel, the president of the Carnegie Corporation of New York. "Carty gave him [Merriam] some excellent advice," Hale later told his wife. According to Hale, Carty advised Merriam to present everyone's opinion to Root, not only his own. Unnerved perhaps by all this attention, Merriam went first to see Rose in his office and assured him of complete cooperation on the telescope project. He said the same thing to Root later in the day. In his report of their meeting, Rose wrote, "Dr. Merriam explained that at the time of our interview in Washington [on April 25] . . . he had had only a very brief conference with Doctor Hale; that they had had no request from the California Institute of Technology; that he had no knowledge in detail of the plan for the telescope and its operation." Merriam was a changed man, or so it appeared.

Rockefeller officials understood perfectly well that Merriam's conversion wasn't worth the paper it was written on. Certain Carnegie trustees, they believed, had "turned the heat on Merriam, insisting that he reverse his position and cooperate." In their eyes, Merriam's capitulation had come about because "of pressure which Hale had manipulated." Carnegie's president knew no more about the 200-inch telescope project going into his second meeting with Rose in New York than he did at the end of his first meeting with him two weeks earlier.

Hale's objective had been "to get the telescope." He got it. In the fall of 1928, the International Education Board of the Rockefeller Foundation gave the green light to Hale's $6 million proposal.

That fall, the Observatory Council, with Hale as chairman, was formed to direct the planning, construction, and operation of the 200-inch telescope. Hale personally assembled the team of scientists and engineers to create it, choosing first John Anderson, a Mount Wilson astronomer, as executive officer. Hale then sent Anderson and his observatory co-worker Francis Pease to Springfield, Vermont, to talk to Russell W. Porter. Porter was an Arctic explorer, artist, and telescope maker. He had designed and built his own observatory and telescope and published many articles on the subject. Anderson and Pease came away from the meeting with Porter convinced he should be invited to join the telescope project. Several weeks later, Porter received a telegram from Hale: "Can you come to Pasadena for several months to assist in designing two hundred inch telescope and instrument shop auxiliary instruments?"

The few months became two decades. Porter arrived in Pasadena on December 1, 1928. He was given the title of associate in optics and instrument design. His first task was to design a small telescope to be used for the site survey for the 200-inch telescope. A dozen such telescopes were made and used for testing sites throughout the southwestern United States in 1929 and 1930. Porter sketched every site he personally visited and, in all, made over a thousand sketches and detailed drawings relating to the project. Palomar Mountain, 160 miles southeast of Pasadena, was selected in 1934 as the site of the 200-inch telescope, and the dome and building housing the great telescope were designed by Porter. "What a privilege to live in an age when such a thing can be done," Rose later told Hale.

The Rockefeller grant also provided for three new buildings on the Caltech campus. Porter completed plans for the astrophysics laboratory first, then turned his attention to the machine shop and the optical shop. By the summer of 1933, all the buildings were in operation. The Palomar team was centered at Caltech, directly involved many Mount Wilson astronomers, sought advice nationwide, and proved both efficient and successful, in spite of its stormy birth.

Meanwhile, Hale had in 1929 also commissioned Elihu Thomson and his associates at the General Electric Company to make large mirror disks of fused silica, which has negligible thermal expansion. They spent more than two years trying, and failed. By the end of 1931, Hale had decided to try somewhere else. He asked the Corning Glass Works to produce a series of Pyrex mirrors—a 30-inch, then a 60-inch, then a 120-inch, and finally a 200-inch. Weight was a major obstacle; it was estimated that a solid Pyrex disk 200 inches in diameter would weigh more than forty tons.

Francis Pease attacked a part of the weight problem, designing a ribbed disk that would cut the weight while preserving the necessary stiffness. Viewed from the back, it resembled a waffle. By June 1932, Corning had made and sent to Pasadena a ribbed-back disk 30 inches in diameter. Following a series of tests in Pasadena, a 60-inch ribbed disk was successfully cast, then the 120-inch; after that, preparations for the 200-inch disk began.

The annealing furnace used to heat the 200-inch disk during the casting operation at Corning resembled a giant igloo. Actually, two 200-inch disks were cast, the second on December 2, 1934. Pouring the molten glass into the mold took seven hours; the disk was then sealed in the annealer, where it remained, cooling slowly, for many months. In January 1936, final plans for transporting the disk to Pasadena were made. It was wrapped in a carpet of felt, cushioned with sponge rubber, and then placed in a steel-plated crate. The crate was then bolted upright in a well-type freight car, built specially for the transcontinental trip by the New York Central Railroad. The disk started its journey west on the morning of March 26, 1936, and the train pulled into Pasadena on April 10. The disk was then hauled to the optical shop at Caltech.

Caltech's 200-inch telescope project brought together a large team of engineers, scientists, and technicians from all walks of life. The chief optician was a man by the name of Marcus Brown, who had started out as a chicken farmer and truck driver on Mount Wilson. Brown had taught himself the optical trade and risen through the ranks of the Mount Wilson optical shop. Bruce Rule, a graduate of Caltech, was the mechanical and electrical engineer of the project. Captain Clyde S. McDowell, the supervising engineer Hale hired in 1934, had taken an extended leave of absence from the Navy Department to come work in Pasadena.

McDowell's responsibilities included the design, manufacture, and construction of the tube and mounting, the grinding table and tools, the dome and its equipment, and the driving mechanisms. Sinclair Smith, a Caltech Ph.D. in physics, class of '24, and a member of the Mount Wilson staff, worked on the controls until he died in the mid-1930s from cancer. He brought to the project, as McDowell later said, "many years of experience as an observer and could see the problems very definitely from the astronomical side and at the same time was a trained experimenter and had had engineering schooling." After Smith's death, Edward Poitras of the Ford Instrument Company stepped in and brought the design of the control system to the manufacturing stage.

By the end of 1936, the mechanical parts of the project were in full swing. Trucks had hauled 5,000 tons of steel up the mountain. Water and electric power were available at the site; a machine shop and housing for observers were under construction; and the steel framework for the big dome stood upright in the ground. Construction work on the tube and mounting had begun at the Westinghouse Electrical and Manufacturing Company plant in Philadelphia.

Many new engineering techniques came out of the project. The shell of the rotating dome, constructed along the lines of an airplane wing, was made of skin plates of steel, three-eighths of an inch thick, butt-welded, with the steel structure supplying the necessary stiffening. Another was the oil-pad type of bearing used to support the weight of the 500-ton telescope. Built in a less health-conscious era, the 100-inch telescope had floated in a pool of mercury. Tests of various design features were carried out on a 1:10 scale model of the 200-inch telescope, later put into service as an operating, student instrument on top of Caltech's astrophysics laboratory. The unconventional tube structure, extraordinarily rigid and lightweight, was designed by Mark Serrurier, a mechanical engineer. His truss design has been adopted worldwide, in all modern telescopes.

The mirror was the heart of Caltech's telescope project. Once

Leading geophysicists met at the Seismological Laboratory in 1929 to chart the future of seismology and to discuss international research collaborations on earthquakes and the interior structure of the earth. *Front row, left to right:* Archie P. King, L. H. Adams, Hugo Benioff, Beno Gutenberg, Harold Jeffreys, Charles F. Richter, Arthur L. Day, Harry O. Wood, Ralph Arnold, and John P. Buwalda. *Top row:* Alden C. Waite, Perry Byerly, Harry F. Reid, John A. Anderson, and Father J. P. Macelwane.

Rocket engine test stand at the Arroyo Seco, in Pasadena, 1936. The now world-renowned California Institute of Technology's Jet Propulsion Laboratory started out as a small group of Caltech graduate students and local amateurs interested in high-altitude sounding rockets. Experimenters at the site include, *from left,* Rudolph Schott; Apollo M. O. Smith; Frank J. Malina, who later became JPL's first director; Edward S. Forman; and John W. Parsons. (Courtesy JPL)

Wilbur Wright, Walter Brookins, Ralph Johnstone, and Albert A. Merrill at the first Harvard–Boston air meet in Squantum, Massachusetts, 1910. Merrill was hired as an assistant in aeronautics at Throop in 1918.

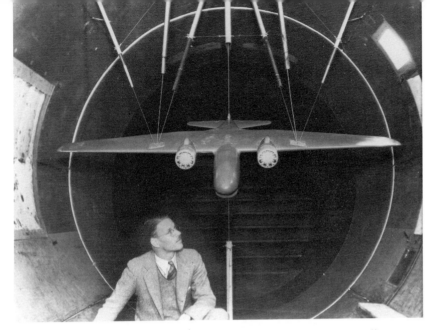

Clark Blanchard Millikan with a model of a DC-1 plane in the 200-miles-per-hour, ten-foot wind tunnel, designed by Arthur L. Klein and Millikan. Completed in 1929, the wind tunnel became a vital tool of the growing aircraft industry. Millikan served as director of the Guggenheim Aeronautical Laboratory from 1949 to 1966.

What happens when an airplane goes to college? This stagger-decalage plane, designed by the Caltech instructor of aeronautics Albert A. Merrill, had no tail and was thus dubbed the Dill Pickle. *Right*, Merle Kelly, the test pilot for the Dill Pickle.

Theodore von Kármán, Hungarian-born engineer and applied mathematician, was hired in 1930 by Robert Millikan as the first director of the Daniel Guggenheim Graduate School of Aeronautics at Caltech (GALCIT). Von Kármán figured prominently in both the rise of Caltech's aeronautics program and the growth of southern California's aircraft (and later, aerospace) industry. In 1962, he became the first recipient of the National Medal of Science, established by President John F. Kennedy to recognize outstanding scientific achievement.

In 1928 the geneticist Thomas Hunt Morgan came to Caltech to organize work in biology. Morgan's fruit fly experiments and the discovery of the chromosomal mechanism by which character traits are passed on from parent to offspring through the interaction of genes earned him a Nobel Prize in 1933.

The Russian geneticist Theodosius Dobzhansky was one of several promising young biologists who joined Thomas Hunt Morgan's *Drosophila* group at Columbia. In 1928, he followed Morgan to Pasadena, where he frequently took advantage of biology's need for live specimens to pursue camping, his favorite pastime.

The physiologist Hermann J. Muller had a desk in Thomas Hunt Morgan's fly room at Columbia from 1912 to 1915. After leaving Columbia, Muller launched his own research team at the University of Texas in Austin, where he became the first geneticist to demonstrate that X rays cause mutations. Muller received the Nobel Prize in physiology and medicine in 1946.

The luncheon party held in the fly room at Columbia University in 1919 to celebrate Alfred H. Sturtevant's return from a brief stint in the army. *Clockwise around the table: Pithecanthropus,* dressed in Sturtevant's army clothes; H. J. Muller; T. H. Morgan; F. E. Lutz; O. L. Mohr; A. F. Huettner; A. H. Sturtevant, holding the beer bottle and a cigar; F. Schrader; E. G. Anderson; A. Weinstein; S. C. Dellinger; and C. B. Bridges.

Russell W. Porter at his drawing board in Pasadena, ca. 1935. Known as an Arctic explorer, inventor, artist, and telescope maker, Porter was recruited in 1928 to help design the 200-inch telescope. Below is a sketch he made in 1929 at Big Bear Lake, one of many sites surveyed for the 200-inch telescope.

The 200-inch disk upon its arrival in Pasadena on April 10, 1936. Scores of people watched the train pull in, including, *far right*, the astronomer Edwin Hubble.

Once in the optical shop at Caltech, the 200-inch disk had to be ground and polished. Shown here is the turntable that held the disk high above the shop floor on a heavy iron frame. Underneath were motors and driving linkages that not only rotated the mirror but also tilted it to a vertical position for testing.

At its dedication on June 3, 1948, Caltech's new 200-inch telescope on Palomar Mountain was hailed as the instrument that "will enable man to undertake his most daring adventure into space." In the years that followed, the Hale Telescope more than lived up to this billing. Its light-gathering power enabled astronomers to look deeper into space (and hence, further back in time) than had ever before been possible and opened previously uncharted regions of the cosmos to observation. Named for the astronomer George Ellery Hale, the instrument reigned as the world's most powerful optical telescope for more than forty years.

Jesse Greenstein with the control box in the Cassegrain cage, located below the 200-inch mirror of the Hale Telescope, at Palomar Mountain. In this chamber, an image-tube spectrograph was used to obtain spectra of faint stars and quasars. The New Yorker Greenstein arrived at Caltech in 1948 to create the graduate school of astronomy that grew up around the Hale Telescope. (Courtesy Caltech Public Relations)

William ("Willy") A. Fowler in 1956 in the W. Kellogg Radiation Laboratory, where he did much of his research in nuclear astrophysics. Fowler shared the 1983 Nobel Prize in physics for his work on nucleosynthesis. (Courtesy Caltech Public Relations)

Charles C. Lauritsen and Robert A. Millikan stand atop the one-million-volt X-ray tube, developed by Lauritsen in 1928. Lauritsen was given the monumental task of directing Caltech's rocket research program during World War II.

Earnest C. Watson, administrative head of Caltech's World War II rocket project, with the Institute's one-millionth rocket. The five-inch high-velocity aircraft rocket, nicknamed Holy Moses, first went into combat use in July 1944.

As famous for his irreverent sense of humor as for being a Nobel laureate, Richard P. Feynman taught physics at Caltech for thirty-eight years. Known throughout the world for his profound mastery of quantum mechanics, Feynman helped create the atomic bomb and shared the 1965 Nobel Prize in physics for unlocking the mysteries of quantum electrodynamics. He died in 1988.

While chairman of the division of physics, mathematics, and astronomy, the physicist Robert F. Bacher initiated construction and use of a new electron accelerator, called a synchrotron, shown here. In 1962, he became Caltech's first provost.

In 1946, the retired chairman of the exeutive council, Robert A. Millikan, turned over the reins to the physicist Lee A. DuBridge. During DuBridge's tenure, Caltech's teaching faculty doubled in number, the campus tripled in size, and new research fields blossomed, including chemical biology, planetary science, nuclear astrophysics, and geochemistry.

the 200-inch disk was safe in the air-conditioned optical shop, the grinding of it into a spherical shape began. As work on the big, rough slab of glass progressed, the work crew uncovered some local imperfections just below its surface, in two spots. The grinding over the entire surface stopped while opticians concentrated on enlarging the blemish into a shallow concave shape in an effort to release stress on the glass. The fracture cleared up as the grinding went deeper, and grinding over the whole face resumed once again. Other fractures cropped up later. Several indentations in the disk's upper surface, which had been caused by the contact of the iron cover with the glass during annealing, also raised concern. To remove the indentations, the opticians ground off more glass from the front of the disk than had originally been planned. The result was a thinner disk, which worried the astronomers more than the physicists, who regarded the additional lightness as an advantage. By 1938, the opticians had removed more than five tons of glass from the twenty-ton blank and had begun polishing it, using a fine abrasive known as a rouge.

From 1928 to 1934, cooperation between the Mount Wilson astronomers and Caltech had been fine. Caltech had paid half the salaries of three Carnegie astronomers—John Anderson, Sinclair Smith, and Francis Pease—from the Rockefeller grant for the 200-inch instrument. Walter Adams, who attended faithfully the meetings of the Observatory Council, had been elected a regular member in summer 1933. Things took a turn for the worse in 1934, however. Early that year, Fred Wright of Carnegie's Geophysical Laboratory prepared at Merriam's request a report about the project. In it, he wrote,

> If the present informal plan [of cooperation] is continued . . . the telescope will ultimately be completed; but C.I.W. will have received little credit, publicly expressed, for its fundamentally important services. Furthermore, with the establishment of a Department of Astronomy at C.I.T., working in cooperation with its strong Physics Department, men from Mt. Wilson will be drawn to C.I.T., if the present "go-getter" policy is still in force. . . . The prestige of

Mt. Wilson will necessarily decrease unless special efforts are made by it to forge ahead in fields not dependent on a 200-inch telescope. ... The 200-inch Observatory will belong to C.I.T. and will gradually be developed into a first rate center for astronomical research. ... The present situation should act as a stimulus to Mt. Wilson to increased activity in the fields for which it is specially qualified.

But Merriam seemed determined not to cooperate in any meaningful way. In 1934, Adams, Mount Wilson's director, was elected to Caltech's board of trustees. Merriam objected. In fact, he now also objected to Adams's serving on the committee that had oversight of the telescope project.

Dumbfounded by Merriam's flip-flop, Hale fired off a salvo of his own. "This great undertaking of the 200-inch telescope, in which the Carnegie Institution is vitally interested, is now menaced by President Merriam's attitude," he warned Carnegie officials in a letter in 1934, pointing out that he, a Carnegie employee, had himself joined the Caltech board in 1907 ("and I have heard of no such objection from that day to this"). Adams picked up his pen too, reminding Merriam that he "had felt no compunction about accepting all sorts of other appointments." Merriam finally caved in and telegraphed his approval to Millikan, but only after Carnegie trustees intervened on Hale's behalf. While Merriam remained in office, no real progress on a joint Caltech–Carnegie Institution astrophysics research program was likely. On this point, Rockefeller, Caltech, and even Carnegie Corporation officials concurred.

In return for the telescope, Caltech had agreed to provide for its maintenance and operation when completed. In fact, the Caltech trustee Henry Robinson had promised in 1928 to provide personally an endowment of $3 million dollars for the telescope. Robinson intended to stand by his promise, but much of his fortune had been eaten up during the depression years. Although the 200-inch project was still going strong when World War II began, none of the promised money was available yet. On the other hand, Merriam had retired and Vannevar Bush had succeeded him at the Carnegie.

Max Mason became the leader of the Palomar project, following his transfer from the Rockefeller Foundation in 1936. He served initially as vice-chairman of the Observatory Council and later became its chairman, a position he held until the observatory and all the work connected with putting it into operation had been completed, in 1950.

America's entry into the war in 1941 interrupted the work on Palomar Mountain. The optical shop continued for a time with the figuring of the 200-inch disk, but work had ceased altogether by the end of 1942. The mirror was boxed, the staff on the mountain reduced to a few caretakers, the astrophysics machine shop converted to war work. The optical shop made mirrors for government wind tunnels, prisms for range finders, and optical parts for various other military devices. Under Mason's direction, the Palomar staff worked on underwater projectiles, the operation of underwater fuses, and other projects for the navy. Not until December of 1945 did polishing of the 200-inch mirror resume. It was completed in October 1947, and after that the mirror was lifted off the grinding machine and placed on a trailer and trucked from Pasadena to Palomar.

The first light ceremony took place in December 1947, with the honor going to the veteran Mount Wilson astronomer and executive officer of the Observatory Council, John Anderson. A colleague of Hale's at the Mount Wilson Observatory going back to 1916, Anderson had overseen the grinding, polishing, and finishing of the 200-inch mirror from start to finish. Using a small, hand-held reading glass for an eyepiece, Anderson peered into the giant mirror. Asked what he saw, Anderson replied nonchalantly, "Oh, some stars."

After subsequent fine polishing in the dome, to improve the precision of the paraboloidal figure, under Bowen's supervision, the 200-inch telescope went into full-time operation in 1949—twenty-one years after Hale had knocked on Wickliffe Rose's door. Neither Hale nor Pease nor Smith lived to see it completed. Bowen, a Caltech physics professor, became director of both the Mount Wilson and the Palomar observatories.

TWELVE

◆

Nuclear Reactions

What we're doing is mainly a cultural and intellectual contribution to the sum total of human knowledge, and that's why we do it. If there happen to turn out to be practical applications, that's fine and dandy. But we think it's important that the human race understands where sunlight comes from.

—WILLIAM A. FOWLER,
1983

AT THE BEGINNING of the last century, scientists took for granted that the sun would shine forever. As the nineteenth century wore on, they spun ever more theories about the source of the sun's energy. One theory held that meteors striking the sun caused it to shine. There were astronomers of that time who believed the sun was fueled by a mass of burning coal. Whatever the theory, it had to accommodate the descriptive sciences of geology and evolution, and Charles Darwin's theory of evolution in 1859 had pushed the earth's age back more than forty million years. By then, the German physicist Hermann Helmholtz had concluded that gravity was a major factor in sustaining the sun's heat. On the basis of his own mathematical calculations, the British physicist William Thomson (later Lord Kelvin) in 1862 put the age of the earth at one hundred million years. By that time, most of Thomson's colleagues doubted that the sun's burning was like the chemical burning on earth.

Following Thomson, the American physicist Jonathan Lane argued that the sun was a mass of gas formed under gravitational contraction, sustained by a centralized fuel supply. In addition, Lane maintained that the sun's mass grew hotter, even while it lost heat through radiation. Lane's contraction theory appealed to many scientists, including Thomson, who believed that a small decrease in the diameter of the sun over the course of a year could provide the sun with sufficient energy to shine brightly. The discovery of radioactivity, just before the turn of the century, breathed new life into the question of what makes the sun shine.

In 1903, John Joly of the University of Dublin and his British colleague George Darwin at Cambridge suggested that radium, if it existed in the sun, supplied a portion of the sun's heat. But if radioactivity was involved, the solar astronomer George Ellery Hale pointed out in 1908, less light would come earth's way ultimately—a possibility "too remote for profitable speculation," concluded Hale.

Profitable or not, the nuclear model of the atom that Ernest Rutherford proposed in 1911, followed by his demonstration in 1919 that the nuclei of chemical elements disintegrate when bombarded by radioactive sources, hammered home the point that radioactivity was a nuclear phenomenon. If Rutherford could transform nitrogen into the heavier element oxygen in his Cambridge laboratory, what prevented the transmutation of chemical elements in the sun? Indeed, in the following year another Cambridge scientist, the astronomer Arthur Eddington, advanced the idea that the sun derives its energy from nuclear fusion reactions within its core. Eddington believed that hydrogen in the sun "combined to form more complex elements," which in turn unleashed a great reservoir of subatomic energy, although he didn't pretend to know under what conditions this energy was liberated in the sun.

Shortly before the start of World War II, several theoretical physicists, including Hans Bethe in the United States and Carl-Friedrich von Weizsäcker in Germany independently advanced a new, more sophisticated theory of how the sun derives its energy from nuclear reactions within its core. They suggested in particular that atoms of hydrogen, the lightest element, could in the sun be converted into atoms of the next-lightest element, helium, by means of a chain reaction involving the isotopes of nitrogen and carbon, which they called the carbon-nitrogen cycle. However, they still didn't know enough to propose a detailed solution to the old question "What makes the sun shine?" In fact, Bethe proposed two possible solutions, the carbon-nitrogen cycle and the proton-proton chain, because not enough was known about nuclear reactions to choose between them. Choosing between them would shortly be the work of the Kellogg Radiation Laboratory, where for six years Charles C. Lauritsen and his student William Fowler, along with others in the laboratory, had been measuring nuclear reactions.

Lauritsen's group had learned about Bethe's work early in 1938 through the theorist J. Robert Oppenheimer, who was then dividing his time between UC Berkeley's physics department and Caltech's. Fowler's reaction to Bethe's *Physical Review* paper on the energy production in stars was instantaneous. "I can tell you," he later said, "it was reading that paper word for word, two or three times, that was the thing that convinced me, boy, this is the way to go!" "For those of us in Kellogg," Fowler later recalled on another occasion, "this was a dramatic event in our lives. What we were doing in the lab had something to do with the stars."

Back in the early thirties, before Bethe's paper launched Kellogg on its astrophysics work, Caltech's Lauritsen, Carnegie's

Merle Tuve in Washington, and Berkeley's Ernest Lawrence were the three big names in American experimental nuclear physics. The three laboratories they directed shared thinly camouflaged institutional rivalries, intense scientific competitiveness, and a healthy dose of respect.

"Lauritsen's qualifications in nuclear physics are obvious," Tuve said of his Pasadena rival in 1937, several years after the three laboratory leaders had carved out rather different research programs. By then, the scientists had largely buried their differences, as well as the urge to rush into print first, right or wrong. Lauritsen had the ability to "work his men very hard and make them love it," he added. Tuve did not exaggerate.

Charles Christian Lauritsen was thirty-four when he came to Caltech in 1926. Born in Denmark, he'd studied physics, sculpture, and architecture at the Danish Royal Academy of Art. In the United States, he worked as an engineer, designing ship equipment, electrical apparatus, and radio receivers for companies in Miami, Cleveland, and Palo Alto. For a while, he was part owner of a radio factory in northern California. In Boston, he owned and operated his own amateur radio station. While working for the Ohio Body and Blower Company, he even invented several measuring instruments. He was making radios for a company in St. Louis in 1926 when he went to hear a lecture by Caltech's head, Robert Millikan. Impressed by Millikan's physics talk, Lauritsen quit his job and drove out to Pasadena with his wife, a radiologist, and their son. He promptly enrolled as a graduate student at Caltech, figuring he would learn "some things in general and especially in physics."

Years later, Charles Lauritsen insisted that he hadn't planned on getting a degree at Caltech. As he once told an interviewer, "I had no intention of staying [at the school] . . . more than a year or so initially. But I liked what I was doing and so I stayed on." Lauritsen was studying high-energy X rays. By the time he received his Ph.D., in 1929, he'd become involved in building a high-voltage X-ray tube for research in both physics and medicine.

Lauritsen's tube could generate over one million volts, and that was the first of many front-page stories about Lauritsen's big X-ray tube in newspapers across the nation. By 1930, he and Robert Millikan had launched a program of research on the treatment of cancer with X rays.

From this work grew the Kellogg Radiation Laboratory. Built in 1931 with funds supplied by the Detroit cornflake magnate W. K. Kellogg, the laboratory led a double life for much of the thirties. By day, Lauritsen's students operated and maintained the high-potential X-ray tube used to treat cancer patients. By night, they did research in X-ray physics. That changed in 1932, when the British physicists John D. Cockcroft and Ernest T. S. Walton showed that it was possible to split atomic nuclei artificially. It was then that Lauritsen modified one of the laboratory's X-ray tubes. He had been using the tube to accelerate electrons in order to produce a beam of high-energy X rays. Save for using an ion source instead of an electron source, Lauritsen had basically the same equipment as the English researchers. Within six months, he had converted one of Kellogg's high-voltage tubes into an instrument with which to accelerate ions and to split nuclei. That step marked the beginning of nuclear physics at Caltech. Now by night, Lauritsen and his students studied proton, deuteron (heavy hydrogen nuclei), and helium-ion interactions with carbon, lithium, beryllium, and boron nuclei, and other chemical elements of low atomic number.

X-ray physics applied to cancer research had seemed to show promise during the 1930s, when medical applications also generated the funds to support Lauritsen's physics research. By 1939, the cancer research treatment had run its course. The doctors and the medical technicians had gone, and the nuclear research that Lauritsen had cultivated and protected during the depression years took over Kellogg.

From the start, the Kellogg group concentrated on the nuclear disintegrations and atomic transformations of carbon and the other light elements. An accomplished structural and architectural engineer, the versatile Royal Art Academy–trained Lauritsen excelled in designing simple, straightforward, and elegant experiments. The group's initial experiments consisted of making transformations from one element to another; to do this, they had to design and build high-voltage accelerators and develop ion sources and detection equipment. Lauritsen's sensitive electroscopes became the industry's standard. In 1934, Tuve challenged in print the results from one of Lauritsen's experiments; Lauritsen simply sent him one of his meters and suggested he measure the rate of radioactivity again. Not long thereafter, Tuve apologized and published a retraction.

Lawrence's equivocal attitude toward Caltech and the Kellogg research group comes out in letters he wrote to Tuve, an old graduate student friend from 1923 at the University of Minnesota. Ten years later, in 1933, Tuve and Lauritsen were in a race to produce artificial neutrons with accelerators. Lawrence knew what was going on at Caltech, because J. Robert Oppenheimer carried the information back and forth between the two institutions. On February 9, 1933, Lawrence wrote to Lauritsen, congratulating him on the production of neutrons: "From Robert Oppenheimer's account of your work, there can be little doubt that you are actually detecting neutrons. I understand also that Tuve in Washington has gotten plenty [of] good evidence of neutrons produced by 600 kilovolt [keV] helium ions." While praising his work, Lawrence wasn't about to concede the definitive discovery to Lauritsen yet.

Lawrence told Tuve as much. In a letter to the Washington scientist dated February 18, nine days later, Lawrence complained about Millikan's influence with the press. "I have noticed a report in 'Science Service' that Lauritsen has produced neutrons, and the usual Cal-Tech ballyhoo is set forth regarding his being the first in the country, etc., to do it." And he urged Tuve to publish his own findings quickly, saying, "Despite the Science Service report, it appears that you are the first one to accomplish it." Tuve knew better. At the end of March, he wrote Lauritsen that he had made an extensive search for neutrons, all "with negative results." Lauritsen, for his part, had achieved positive results, but he refined and repeated the experiments again and again before sending the paper announcing his discovery of artificial neutrons to *Physical Review* in September of 1933.

The historical connection between the experimental work in Kellogg and Bethe's later deduction that fusion powers the sun and the stars turns on a discovery made by Lauritsen and his graduate student Dick Crane in 1934. Lauritsen and Crane bombarded carbon 12, the most abundant of the carbon isotopes, with protons; to their surprise, the nuclei did not disintegrate—or so it seemed. When the proton-bombarding energy exceeded 650,000 electron volts, the two observed radioactive nitrogen 13 and gamma radiation instead, suggesting that their target had not lost particles but had instead acquired new ones: had undergone not fission but fusion. However, Lauritsen waged a lonely battle in the beginning; few physicists, Oppenheimer included, were prepared to believe in 1934 that a particle could be added to a nucleus without some other particle's being spun off to carry away the excess energy. Tuve, in fact, had quarreled publicly with Lauritsen's findings, attributing his observations to natural deuterium (heavy hydrogen) contamination. When Lauritsen pressed him to repeat the experiment, Tuve replied, "OK, Charlie, lend me one of your electroscopes." This time, Tuve and his collaborator Lawrence Hafstad also found the telltale narrow peaks, called resonances, in the excitation curves for the proton-induced activity.

The proton-carbon excitation curves bore little resemblance to the smooth and continuous reaction rate curves associated with deuteron energy. "This marked difference in excitation curves," Fowler later wrote, "convinced Lauritsen and Crane that protons did indeed produce nitrogen 13 in carbon bombardment." The reaction Fowler described is an example of a process known as radiative capture. In this process, the projectile is captured by the nucleus and forms an excited state of a new isotope, or element, which then decays to its ground state by radiating a gamma ray. It is "resonant" because the projectile must have just the right energy to form the excited state of the new species. Carrying out the proton-carbon reaction is a little like flipping an acrobat up from a teeter-totter, so that he winds up standing on the shoulders of his fellow performer. He must be given almost exactly the right push, or he won't wind up in the right place. Similarly, the proton, if it has precisely the right energy, can be captured by the target carbon 12 nucleus—not on its shoulders but in an excited, or energetic, state of a new combined nucleus, nitrogen 13. Then, just as the acrobat must eventually come back to the ground, so does the nitrogen 13. By giving off a gamma ray (ultrahigh-frequency light) it sheds its excess energy and falls into its normal, low-energy state, which is in fact called the ground state by physicists. Unlike the acrobats, however, the nucleus doesn't divide back into the original two pieces when it drops back to the ground. Instead, it falls into the ground state of the combined nucleus, nitrogen 13.

The discovery of radiative capture of protons by carbon set the focus of Kellogg's nuclear physics program for the rest of the decade. Convinced that the excitation levels in the light nuclei were the key to understanding the structure of the nucleus, Lauritsen and his students undertook detailed measurements of nuclear reaction rates of all the light nuclei. Fowler received his doctorate in 1936, and for the next three years he and Lauritsen spent much of their time studying excitation curves, the yield of the activity produced versus energy, for the carbon and nitrogen isotopes bombarded with protons.

At the end of the decade, Bethe made his suggestion that the thermonuclear reactions underlying the conversion of hydrogen into helium in the stars depends in a crucial way on the catalytic process known as the carbon-nitrogen cycle. In the first of the six nuclear reactions involved in the transformation cycle, a nucleus of carbon 12 fuses with a proton or hydrogen nucleus to yield a nucleus of nitrogen 13 and a gamma ray. Bethe's first proposed step in the cycle matched exactly the reaction that Lauritsen and his colleagues had produced in Kellogg in 1934. Although the scientists in Kellogg couldn't know it at the time, this capture-and-combine process was typical of a long and complex series of reactions that govern the burning of the sun. Ultimately, to explain quantitatively how the sun works, all the energies and all the capture probabilities of the light elements would have to be measured and graphed as excitation curves. However, in 1934, the reason for measuring excitation curves was not yet to explain how the sun works.

The first job was to convince the scientific world that nuclei could do these seemingly unlikely acrobatic tricks. Physicists such as Tuve and Oppenheimer were skeptical or outright hostile. But even beyond proving the existence of these precisely defined states, which showed up as sharp peaks on the excitation curves, there was the matter of finding more of them and measuring their height and their energies. The very existence of something precise to measure in the otherwise formless excitation curves appealed to the scientific instincts of Kellogg's physicists.

"When Bethe came out with the carbon-nitrogen cycle, we kind of felt a proprietary interest in this group of reactions," Fowler recalls, "because we had been working on them. . . . [It] all tied very closely together." By 1939, the Kellogg researchers had switched from an alternating-current high-voltage tube to a two-million-electron-volts direct-current Van de Graaff electrostatic accelerator, capable of high-resolution work. With the new Van de Graaff machine, Lauritsen and Fowler had begun to measure very carefully all the effects associated with resonance phenomena. Experimental work on reaction rates at resonance, locations and widths of resonances, and gamma-ray spectra at low energy in the light elements boomed. For the first time, Fowler and Lauritsen were in a position to do very careful excitation curves—and only by accurately measuring nuclear reaction rates could problems such as Bethe's application of nuclear physics to astronomy be solved.

Pearl Harbor abruptly ended all such work in Kellogg's peacetime research program, and with it any immediate chance for Lauritsen and Fowler to follow up Bethe's theoretical ideas. After the war, however, Fowler, Lauritsen, and others took up the task again. Postwar strategies for studying thermonuclear processes in the stars included a series of informal, weekly seminars with Mount Wilson astronomers at Director Ira Bowen's house, and Fowler's collaboration with a diverse group of scientists ranging from the cosmologist Fred Hoyle to the astronomers Margaret and Geoffrey Burbidge. In 1948, Jesse Greenstein came to Caltech to organize work in astronomy, and his interests, particularly in the abundances of the elements in stars, stimulated Caltech's nuclear physicists to pay more attention to the astronomical side of nuclear astrophysics. However, the most important step was to initiate an experimental program that would strike at the heart of Bethe's theory. In 1946, Fowler's graduate student R. N. Hall took as his topic for a Ph.D. thesis the determination of the rates of the reactions in the carbon-nitrogen cycle at stellar conditions. Four years later, Fowler and Hall published their first paper on the problem.

Hall's problems were considerable. First, it took time to build a low-energy accelerator to simulate low, stellar energies. Even so, the terrestrial laboratory energies were too high, and extrapolation to lower energies was unavoidable. Moreover, the reactions that Hall was after occur infrequently at low energies. Indeed, it took Hall three years just to accumulate a thousand nuclear events. In the sun, the effective energy for the carbon-proton interaction measures only 30,000 electron volts; the machine Hall built in Kellogg was a low-energy, 150,000-electron-volt machine—this was the lowest energy the physicists could get in the laboratory and still detect something. Ironically, in the end, Fowler and his students concluded that the carbon-nitrogen cycle is not the dominant process in the sun. Kellogg's own reaction was not the key to solar energy after all.

To be sure, Bethe had also suggested another process, the

proton-proton chain. The measurements made in Kellogg supported the latter process, in which protons combine to form helium, with the emission of large amounts of energy. To the question "What does the sun shine on?" Fowler's group answered, "It starts with the proton-proton chain." That answer launched the science of experimental nuclear astrophysics.

THIRTEEN

◆

The Rockets' Red Glare

War does upset plans, doesn't it?
> —FRANK B. JEWETT, 1942

... I still think that we should try to make the Institute as useful as possible, partly for the sake of the Institute and partly because our facilities should be a great asset in any important development work.
> —CHARLES C. LAURITSEN, 1940

My own philosophy has been that since the main function of the federal government is defense, it is clearly correct to take federal funds for Army and Navy research or for any research which can be immediately classified as an essential part of our national defense.
> —ROBERT A. MILLIKAN, 1945

ROBERT MILLIKAN had impressive industrial, financial, academic, and government connections. Without them, Caltech could not have survived the 1929 stock market crash and the Great Depression, which settled over America in the thirties.

In spite of the failure in 1930 of the trustee Arthur Fleming's large trust, Millikan balanced the budget and found money enough to support special research projects.

The faculty met Millikan halfway, voting to take a 10 percent cut in salary in 1932. Millikan was counting on this gesture to meet half of the school's deficit.

During the depression years, Millikan, George Ellery Hale, and Arthur Noyes personally met a small portion of Caltech's deficit. For a time, for instance, Hale and several trustees of the Institute supported Linus Pauling's chemical research. To make up the balance of the deficit, which in 1932 approached $80,000, Millikan spoke to several key friends of the Institute. Each friend had already pledged money for a particular building. Millikan asked if the income from these funds could be diverted to the meeting of Caltech's current expenses. The donors agreed. Aside from the 200-inch telescope project, building construction on the campus came to a halt. It resumed with gusto five years later, in 1937. The Arms and Mudd Laboratories of the Geological Sciences began to take form to the west of the Astrophysical Laboratory. Across the cypress-flanked mall, behind Gates Chemical Laboratory, the excavation for the Crellin Laboratory of Chemistry and the second unit of the Kerckhoff Laboratories of the Biological Sciences also began.

Millikan consulted his faculty less and less during the thirties. No one challenged him. His considerable fame was raising a lot of money for the Institute. Even so, not all of the faculty were pleased. Neither were the students. An anonymous Caltech graffiti artist added his message to others on the steam shovel several days after it arrived on the campus to begin excavations for Crellin. Underneath "Roosevelt for King" and "Jesus Saves," the student wrote "But Millikan Gets Credit." Students often have the last word at Caltech.

When war broke out in Europe in 1939, Millikan hoped to

be called to Washington to direct the mobilization of civilian scientists for national defense, as he had done once before, in 1917. The call never came. It went instead to the Carnegie Institution's president, Vannevar Bush. In the late spring of 1940, the MIT-educated electrical engineer and inventor of a mechanical computer known as the differential analyzer assumed the chairmanship of the new National Defense Research Committee (NDRC). Created by President Franklin Roosevelt at Bush's behest, the NDRC was a federal agency with the power to award defense contracts and conduct research on military weapons. Its purpose was to help the army and the navy on research problems. Eager to match research with military needs, the NDRC surveyed the scientific talent and laboratory facilities available in universities across the land.

Caltech had already done this on a small scale. At the urging of the faculty board's chairman, Earnest Watson, the school had in spring of 1940 formed the Council on Defense Cooperation, under the direction of Richard Chace Tolman, dean of the graduate school. Max Mason, chairman of the Observatory Council, served on the new council along with Watson and Tolman.

All three men had experience preparing for war. Mason had gained his experience at the Naval Experimental Station in New London, Connecticut, during World War I. A mathematics professor at the University of Wisconsin, Mason had demonstrated a flair for experimental work at New London by developing an acoustic system for detecting enemy submarines. Watson had been working on a Ph.D. in X-ray physics under Millikan at Chicago when World War I came along. Eager to see action in Europe, he had enlisted in the navy, at Millikan's urging, and had wound up working on the antisubmarine problem at New London too, along with Mason. Watson didn't get to go overseas; he was the only enlisted man among the scientific group there, and he was constantly seasick. "If I had played it right and I had gone into the navy as a commissioned officer, I could have done much more," Watson griped years later, still blaming Millikan for not having told him to apply for a commission. Soured by his own earlier experiences in dealing with the navy, Watson was ready in 1939 to serve his country again, but determined not to make the same mistake twice.

Tolman's wartime experiences had been very different from Watson's. In the closing days of the war, Tolman had resigned from the faculty of the University of Illinois to accept the rank of major in the army and to serve as chief of the newly established dispersoid section of the Chemical Warfare Service. Charged with studying the production of toxic and nontoxic smoke screens and candles, the division also tested airplane ammunition, using the "hangfire measurer," a quality-control machine developed by Tolman. When the Fixed Nitrogen Research Laboratory of the Department of Agriculture began operations, at the close of the war, in the old headquarters of the Chemical Warfare Service at American University, Tolman had plunged wholeheartedly into his new responsibilities, first as associate director and then as director. He once described the place as "a great mixture of business, science and politics" and added, "I enjoy all of them." Afterward, Tolman had joined the faculty of Caltech as professor of physical chemistry and mathematical physics.

Tolman's intellectual and political clout on the Caltech campus was considerable. In explaining why he had appointed Tolman defense council chairman in 1940, Watson later said,

> I had to set it up in such a way as not to have Millikan take it over; so I had to get Tolman and Mason who were both members of the Executive Council [Millikan was its chairman] on it. They were natural people on the committee. Tolman . . . was one of these people who had a great sense of his own importance and could only function if he was in charge of it, so Tolman became the first chairman of the committee.

Greener pastures beckoned Tolman. He headed for Washington at the end of the spring 1940 term, leaving Watson to do most of the defense council's work anyway. By the time Watson had finished compiling the "Report on Possible Contributions of

CIT to the Problem of National Defense" in mid-June, Tolman had taken up residence in the Hay-Adams House in Washington and been appointed Bush's vice-chairman and head of the NDRC's division dealing with armor and ordnance. He meant to stay there for as long as necessary, "as long as the emergency lasts," he told Millikan. Tolman didn't return to Pasadena until the war ended in 1945.

The Caltech nuclear physicist Charles Lauritsen, "Charlie" to his friends, headed the list of scientists Tolman in turn meant to recruit for national defense work. "I want to know if you would consider the possibility of coming on here if the right opportunity should develop," Tolman wrote Lauritsen, at the end of June of 1940, adding, "You would have to explore this very cautiously since I presume the Institute does not want to lose its top men even temporarily. . . . Your brains and sense would make a big dent in Washington. You might even bring on your whole crew. . . . Please write me. I miss your counsel and advice every day."

In his reply to Tolman's "Dear Charlie" letter, Lauritsen pushed for using Caltech as a base of operations but did not close the door on Tolman's Washington offer. Tolman tried again, this time appealing to Lauritsen's innate talent and fondness for designing and building all kinds of things from scratch. He also promised Lauritsen a minimum of red tape, for

> these good contacts with the Army and Navy mean, in the first place, that we can really find out whether a proposed device for investigation would have any military usefulness if it could be developed . . . in the second place . . . the Army and Navy themselves propose devices on which they would like help. . . . I certainly think that you ought to come on and help, at least while things are getting started.

To sweeten the offer, he also dangled a round-trip ticket in front of Lauritsen:

> Later on you may then want to go back to Pasadena, and will be in a position to take charge of really useful military investigations which can be done on the coast. . . . I think that Dr. Millikan will also see

the importance of making direct contacts with Washington in order that CIT may be able to do its share to help in the emergency.

Lauritsen gave in. Appointed Tolman's vice-chairman in the armor and ordnance division of the NDRC, he assumed his new duties in Washington in August 1940, leaving open the question of when he would return to Caltech.

Millikan took a dim view of Lauritsen's request for an indefinite leave of absence. In part, he thought it "at least possible" that Europe's war would be resolved within a year. In that case, he told Lauritsen, it made more sense to work on short-term military projects that had a foreseeable payoff than to undertake projects that required several years of development. In part, Millikan believed that national defense projects could be carried out more effectively in university laboratories and in Caltech's, in particular, than in Washington or someplace else. As he wrote to one of his co-workers, who was thinking of leaving Caltech to join the physicist Lee DuBridge's large-scale radar project at MIT, "I am convinced, as is everybody here who has had a chance to express himself, of the mistake which is being made in trying to concentrate fifty prima donnas in physics at any one spot." Millikan went on to recall some of his own wartime experiences:

> [Max] Mason and I, who have talked it over and who had intimate experience with this sort of thing in the last war, are of one mind about it. At New London we never got more than three or four prima donnas on one problem together, and Mason says that he had to send back one or two of the men whom he had pulled in from Madison to their own laboratories to get things done rapidly. . . .

More than anything, Millikan feared that he would not be able to keep Caltech's own physics faculty intact unless he had a suitable defense project anchored at Caltech. His best chance for getting one rested with Lauritsen.

Lauritsen saw to that. Work on the design and development of rocket projectiles began at Caltech's Kellogg Radiation Laboratory under his direction on September 1, 1941, three months before the Japanese attacked Pearl Harbor and America entered the war, and did not stop until September 1945. From a modest government contract—the initial allocation amounted to $200,000—in 1941, Caltech's rocket project grew into an $80 million war industry. In 1944, at the height of the project, Caltech was spending $2 million a month on the work—its annual prewar budget had been $1.25 million. By 1944, the one-millionth rocket, a five-inch high-velocity aircraft rocket, known as the Holy Moses, rolled off the assembly line in a 38,000-square-foot plant on the outskirts of Pasadena. No small feat, William Fowler later remarked, adding, "But the upshot was that a large part of Caltech literally became a branch of the Bureau of Ordnance." By then, the number of men and women on Caltech's payroll had climbed more than tenfold, to nearly five thousand.

Rockets are among the oldest weapons of war. As military weapons, they were used by the Chinese against the Mongols during the thirteenth century, by the natives of India against the British in the eighteenth century, by the British against the French and the Danes duing the Napoleonic Wars in the early nineteenth century, not to mention the British against the Americans during the War of 1812. On the last occasion, the enemy's rockets scattered the American militiamen guarding Washington, permitting the British to enter and burn the young nation's capital. Soon afterward, the British fleet attacked Baltimore, firing rockets from ships in the harbor. During that attack, which failed, an American poet and lawyer by the name of Francis Scott Key wrote "The Star Spangled Banner." A famous couplet of the national anthem, "the rockets' red glare, the bombs bursting in air," sings of what Key himself saw that night. By the Civil War, large mounted guns had replaced rockets on America's battlefields. Rockets continued to be used for signaling, but for little else.

Modern warfare found new uses for rockets. Regardless of their size and weight, all rockets consist of a payload and a motor containing the propellant powder. The rockets of World War II destroyed tanks, blew up aircraft, and attacked submarines. Servicemen aimed rockets against military targets from fighter planes, as well as from landing craft as they approached shore. In the closing days of the war, the Germans fired long-range V-2 rockets against British civilians.

Isolationism, a legacy of nineteenth-century American history, hindered President Roosevelt's ability to organize the country's defense needs during the thirties. The U.S. ambassador to England, Joseph P. Kennedy, spoke for many of his countrymen when he said in 1938, "I can't for the life of me understand why anybody would want to go to war to save the Czechs." When Germany invaded Poland in August 1939, Congress reluctantly repealed the arms embargo. It was only in the following September, after Hitler's armies swept over western Europe, occupying Denmark and Norway in April, Holland and Belgium in May, and, finally, France in June of 1940, that the president told the American public that he had pledged American destroyers and other aid to Britain in return for leased bases in the Caribbean. Vowing to transform the country into a "great arsenal of democracy," Roosevelt publicly announced in November that the United States would lend arms directly to Britain as well as give outright half of its own future weapons production. Roosevelt had ships, planes, and tanks in mind, not rockets.

America had no rockets in its ordnance depots, nor was it likely to get any, as Charles Lauritsen quickly discovered when he came to Washington in the summer of 1940. At the request of the Naval Bureau of Ordnance, Lauritsen's NDRC group, known as Section L (the L stood for Lauritsen), began working on the development of proximity fuses for bombs, rockets, and artillery shells. Unlike fuses that work by impact or by time, the proximity fuse detonates when it is near the target—making it a far more lethal antiaircraft and ground weapon. Early studies on different types of proximity devices included work on fuses triggered by sound, radio reflections, optical sighting, and infrared emissions. In January 1941, William ("Willy") Fowler, Lauritsen's right-hand man in the Kellogg Radiation Laboratory, packed his bags and rode the train from Pasadena to Washington, accompanied by J. Streib and R. Becker, the newest Ph.D.'s in the Kellogg research group, leaving behind three graduate students. Fowler and the others started off working on photoelectric proximity fuses at the Carnegie Institution's Department of Terrestrial Magnetism. Another Kellogg nuclear physicist, Thomas ("Tommy") Lauritsen, Charles Lauritsen's son, soon joined them; he also worked on proximity fuses at the Bureau of Standards.

It didn't take Charles Lauritsen long to decide that the country needed rockets more than proximity fuses. Fowler's own experience helped turn Lauritsen into a rocketeer. Fowler's first assignment had involved working "on proximity fuses for large bombs to be dropped from one airplane down on another, and as the bomb went by the target airplane, the proximity fuse would trigger." It wasn't necessary to hit the target. To test them, the devices were attached to rockets under development by the navy at Indian Head, Maryland. But Lauritsen's group didn't often get to find out whether their devices worked, for as Fowler recalled, "We found that the rockets that they [the navy] were using were quite unreliable, and the usual result was that the rocket would either blow up on the launcher and completely demolish our hand-built proximity fuse, or the thing would take off and just plop in the Chesapeake Bay." Lauritsen blamed the solvent in the propellant powder used in the rocket motors fired at Indian Head, pointing out that the British-made 3.25-inch-diameter antiaircraft rocket employed a powder that was manufactured differently.

The British used solventless sheet powder; Lauritsen wanted the rocket engineers in charge of the NDRC's Maryland project, Dr. Clarence Hickman and Army Major Leslie Skinner, to develop a similar powder. They declined to do so. "It was claimed in this country," Thomas Lauritsen later said,

that this was impossible to do. The only rocket propellant effort that was going on at the time was an effort at Indian Head in which the primary thrust was wet extrusion in which one dissolved a material in a solvent and extruded it wet and then allowed it to dry. In the course of drying, of course, it got badly out of shape; it got bubbles, and was really a terrible powder in all respects.

By means of this solvent-extrusion method, only small grains of powder could be produced. And small (thin-web) grains, while suitable for small-arms ammunition, sharply limited the range and the velocity of the rockets they propelled.

The mass of extruded powder (the grain) had another problem, as Lauritsen's son went on to explain:

> Probably the most important difficulty, though, was the web was so thin. [The thickness of the web regulates the amount of time it takes to burn the grain.] It took weeks to dry even a ¼-inch web and one wanted to go for 1-inch or more. Charlie was very impatient about that development. Here we had fuses but not rockets, so he started up the rocket project out here [at Caltech] and sent Fowler back to get the things organized.

As Charles Lauritsen learned when he visited England in the spring of 1941, British powder manufacturers had the experience and heavy machinery required to make large, long-burning, thick-web rocket powder. They did this by a process called "dry extrusion." Without using a solvent, they mixed the dry propellant ingredients and rolled them into a sheet, wound the sheet powder into "carpet rolls," and squeezed them through a die under high pressure.

The powder problem wasn't Charles Lauritsen's only headache in 1941. "There is, as you know, a strong tendency to keep all development work concentrated here in Washington as far as possible, and the groups working here are often very reluctant to let go, especially of anything good," he complained to another Caltech professor that spring. By then, Lauritsen was betting that military rockets would play an important role in the war, "at least until more and better guns can be produced which will not be soon." But he couldn't get Hickman (a rocket pioneer who had served as Robert Goddard's assistant in 1918–19) and Skinner (another great believer in military rockets, who had performed numerous experiments in the 1930s on his own) to see things his way. In April, they refused to expand the rocket work at Indian Head (they were then working on armor-piercing bombs using rocket projectiles), and they balked at encouraging the NDRC to support work on rockets elsewhere. Taking matters into his own hands, Lauritsen wrote to Tolman in April, urging him to sponsor an expanded rocket program without delay. If it was up to him, added Lauritsen, he would develop different types of military rockets, starting with a high-altitude aircraft rocket and a plane to plane rocket. Caltech, he pointed out in passing, was well qualified to tackle the aircraft rocket project.

Tolman was less than thrilled by Lauritsen's request. "I do not find it easy to take an objective view with regard to a plan that might tend to decrease the time Charlie spends in Washington where he is so badly needed," he cautioned Bush, passing the letter up the line.

But Lauritsen saw himself playing a less and less useful role in Washington. The desire to return to Pasadena grew stronger by the day. On July 30, 1941, shortly after he was appointed a member of the NDRC's Indian Head committee, Lauritsen tried again to get the Maryland rocket group to expand its work. According to notes he made at the time, Hickman and Skinner "felt that they had neither responsibility nor authority to make recommendations for expansion." Although they had initially opposed Lauritsen's plane-to-plane rocket, work on such rockets was now under way at Indian Head, in addition to work on an armor-piercing bomb. According to Lauritsen, Maryland's rocketry group "also felt that other rocket work would be hopeless for a long time to come due to the fact that the necessary thick webbed powder was unobtainable in this country." Worse yet, Skinner was of the opinion "that neither the Army nor the Navy would be interested [in high-altitude rockets] as they could be obtained from England if desired." For Caltech's defense-minded nuclear physicist, this was the last straw.

In a lengthy report addressed to Vannevar Bush, director of the government's Office of Scientific Research and Development, on August 1, 1941, Lauritsen summarized the rocket situation in America and in England and called for an immediate expansion of rocket development at home. Having visited England several months before and studied its rocket operation, he was more convinced than ever that rockets would play a key role in World War II. "Does the Army and Navy look to us to keep them posted on rocket developments and to make recommendations regarding promising developments or do they prefer to make such investigations themselves and suggest projects to us?" Lauritsen asked Bush.

Bush answered the question by urging Lauritsen to form a new rocket section at Indian Head under Tolman. Lauritsen refused. Hickman, he countered, should enlarge the membership of his own rocket section, and he recommended Watson and Warren Weaver as new committee members. Bush, who felt strongly the need to expand the country's effort in rocket research and development ("as promptly as possible and in the particular locality and under the particular auspices which seem to be most effective after a full study," he told Tolman at the time) was miffed. Charles Lauritsen later wrote, "Dr. Bush accepted this suggestion most reluctantly, stating that he did not like it and that he would hold me personally responsible for our Rocket program."

Tolman meanwhile had come around to the idea of contracting for defense work on rockets in southern California under Lauritsen's direction. Indeed, Tolman and Lauritsen had already talked with Millikan in Pasadena during the summer about starting up such work at the school. In a letter dated August 9 and hand-delivered to Bush by Lauritsen himself, Tolman now advocated that Bush support work on a high-altitude rocket at Caltech, and he suggested a $200,000 budget for the first year's work.

When Hickman's enlarged section (Section H) met on August 19, Skinner informed the group that the army was anxious to have an antiaircraft rocket in its arsenal, so much so, in fact, that it didn't want anyone else to develop it. Skinner and Hickman also objected vehemently to giving a $200,000 rocket contract to Caltech. If anyone deserved a contract, they said, it was Robert Goddard, the reclusive and secretive American rocket pioneer. The discussion grew heated. Several committee members, including Earnest Watson, raised their hands. The contract under discussion, they pointed out, "could not in any sense be considered a plum and . . . it should be placed where the work could be carried out most effectively." Nevertheless, no action was taken on the Caltech contract.

Lauritsen could be very persuasive. Bright and early the next morning, he met with General Joseph Green, chief of the Coast Artillery, and Colonel Chamberlain. Green and Chamberlain were interested in a moving target for training gunners. What did Lauritsen think of using a rocket to tow a target? General Green asked. Lauritsen urged him to consider using a rocket with large fins for target practice instead. Green's eyes lit up. If such rockets were available immediately, he could use a thousand a day on the firing range, he told Lauritsen. While the target rocket wasn't a rocket for use as a weapon, Lauritsen's lobbying had purchased him the military backing he needed at that particular moment. In speaking of his father's uphill battle to corral support for an expanded rocket program before Pearl Harbor, Thomas Lauritsen remarked, "I think Charlie worked much more with individuals than through organizations. I guess there isn't any other way of doing it, but one thinks of writing a letter, 'Dear Bureau of Ordnance—I have the following': But Charlie didn't do that. He kept going around until he got hold of somebody whom he could understand and who could understand him—then things got done."

Indeed, within the week, the full National Defense Research Committee had earmarked $500,000 for a general expansion of rocket work, Hickman's section had reconsidered, accepted, and recommended Caltech's rocket proposal to the NDRC, and Tolman had sent word to Millikan that the defense contract they had talked about was in the works. "Charlie had always been a very forceful man," his longtime colleague William Fowler once said, "and you either liked him very much or perhaps not at all." Lauritsen's own research team stood behind him, 100 percent. "We did what Charlie told us," Fowler recalled, adding, "If he had told us to stay working on the Indian Head propellant, we would have done it." Charlie told them to pack their bags and return to Caltech instead.

Director of Research C. C. Lauritsen's Caltech rocket program got under way on September 1, 1941. The contract covered "development and testing of rocket devices" in general, the "development, adapting and testing" of fuses, projectors, and other related equipment, and "such other development work in the field of armor and ordnance as the Committee may request." In time, the navy became Caltech's best rocket customer.

Strictly speaking, no rocket work should have been started until February 1942, when officials back east officially approved the contract, but Robert Millikan and Lauritsen and the trustees all agreed on the importance of going ahead with the work without delay. Backed only by a letter of intent from Washington, the trustees guaranteed the necessary money, so Lauritsen could start putting together a team immediately. From Caltech, he recruited the chemical engineers Bruce Sage and William Lacey, the physicists Ralph Smythe and Carl Anderson, and the electrical engineer Fred Lindvall, among others. He also tapped a number of Mount Wilson astronomers, including Ira Bowen. Millikan seemed pleased by the turn of events. He "approved hook, line, and sinker," as Fowler put it, of Caltech's rocket project.

On the other hand, Robert Millikan was seventy-three years old, too old to come to grips with the sheer magnitude of the rocket project. Indeed, he had very little to do with the project directly. Earnest Watson, who had originally come out to Caltech to get Millikan's physics department started after World War I, served as the project's official investigator. This meant, among other things, that Watson approved purchases, leases, and expenditures as they came up, not Millikan. Having done all the administrative work for Caltech's physics, mathematics, and engineering division for many years, Watson was well prepared for his new administrative duties. He also understood what was expected of him:

I didn't take the leadership; my job was to make it possible for Lauritsen and Fowler and . . . these other people, to do their job. You had to know how these people worked so you could take an individual and allow him to make his contribution. Sage who made an enormous contribution to the explosives end of the game . . . [was] a very difficult person to work with. He and Lauritsen were much of the time at each other's throats, and Sage was terribly independent. He would have just pulled out if I hadn't smoothed things over all the time. My job was in the first place keeping Millikan out of Bush's hair, and Millikan out of Lauritsen's and the group's hair so they could go ahead and do something without having to go through Millikan at every turn.

If Watson had crossed Lauritsen, Lauritsen would have had the last word. If he had crossed Millikan, Millikan would have had the last word. Watson never crossed either of them. But he paid a high price. "Subconsciously I just put this whole thing out of my mind," he once said many years later. "If I'd kept remembering it, I'd get upset and under tension and couldn't do anything else. . . . A lot of my memory on World War II is not very specific because I've forgotten; I've purposely wanted to forget."

Watson was the perfect foil for Lauritsen. Recalls Fowler, "Earnest was just the opposite to Charlie Lauritsen in that he was able in his very quiet and amiable way to keep the offices in Washington pacified even though Charlie was sending off letters saying that things technically weren't all that they ought to be." They were the perfect odd couple.

One of the things that Lauritsen kept complaining about was the powder problem. In October of 1941, he couldn't get enough powder, not even solvent-extruded sticks. In addition, the quality of the materials that the group did receive was poor. Out of desperation, Lacey, Sage, and Lauritsen decided to design and build their own extrusion press. By late fall, C. C. Lauritsen reported success on this front:

> In about two weeks and a cost of a few hundred dollars a very crude and temporary press was constructed and put into operation in Eaton Canyon. After a few days of experimental operation the work was turned over to two inexperienced laborers who operated the press without the slightest difficulty and extruded all the available sheet stock. There has been no evidence of defects in the powder produced from solventless sheet by this method, either by inspection, partial burning tests, or actual firing.

While Lauritsen savored the Eaton Canyon triumph, he also berated Tolman for not doing enough for Caltech's rocket project. "By operating the presses one day a week we can extrude all the powder needed for experimental purposes as soon as sufficient sheet material becomes available," he wrote to his boss on December 18, 1941, eleven days after Japan bombed Pearl Harbor. "There is, however, little likelihood that a sufficient supply will be made available to us unless we succeed in gaining more active interest in and support for this work from high ranking Army and Navy officers. I am confident that the lack of interest is due largely to lack of information regarding the possibilities of rockets." Two thousand pounds of sheet ballistite (the nitroglycerin- and nitrocellulose-based propellant used in trench mortar shells) arrived in Pasadena early in February 1942, enough for experimental purposes.

Located in the foothills just north of downtown Pasadena, the Eaton Canyon powder plant grew by leaps and bounds during the war. The first 6 acres of land were leased from the city in fall 1941. Within a year, an additional 140 acres were required for the extrusion presses and other equipment. Concrete barricades went up around the facility as a safety precaution. It was Bruce Sage's operation, and he ran the propellant plant around the clock, twenty-four hours a day, seven days a week. With Fowler and his wife, Ardie, Sage organized the work force, all amateur labor, into three eight-hour shifts—a day shift, a night shift, and a swing shift. Fowler recalls them "filling powder bags to ignite rocket propellant." Like the rest of the Caltech rocket project, Sage's powder factory was a classified operation. "Very few people realized we had enough high explosives up there to have blown Pasadena off the map," Watson later recalled.

The rocket project took over the Kellogg Radiation Laboratory completely, spilling over into the optical shop, the steam plant, the astrophysics shop, Bridge Laboratory of Physics, the Astrophysical Laboratory, and several dwellings adjacent to the campus as well. At the height of the project, Caltech subcontracted with three hundred shops in the Los Angeles area for the fabrication of various rockets parts—nozzles, firing plugs, motor tubes, and nozzle plates. Proving grounds ranged from the Goldstone Testing Range, on the Mojave Desert, to Camp Pendleton, to the naval air station in San Diego, and to Morris Dam.

Rocket devices designed and developed at Caltech during World War II included target rockets used for training antiaircraft gunners, barrage rockets on small craft used in landing operations, airplane rockets, packing shells ranging from 3.25 inches in diameter to 5.0 inches, and a rocket-propelled antisubmarine bomb. The antisubmarine rocket was the first Caltech rocket to be used against the enemy.

Enemy submarines were in the fall of 1941 very much on the mind of Max Mason, Caltech's World War I antisubmarine expert. Mason organized and directed a series of studies on the underwater ballistics of projectiles at Morris Dam, located seventeen miles from Pasadena, near Azusa. Eager to develop a rocket-projected antisubmarine bomb, Mason turned the problem over to Lauritsen's rocketeers, who, in turn, made up some sample bombs weighing close to forty pounds. These were successfully fired early in 1942 by means of rocket propulsion. By then, German submarines along the Atlantic seacoast and in the Caribbean had become the biggest threat to the United States. When suitable powder became available later in 1942, work on an antisubmarine rocket finally took off. Eaton Canyon reverberated with the firing of test models. Other tests were carried out on the Mojave Desert. The projectile was tested at sea, off the coast of San Diego, three months later, on March 30. As Charles Lauritsen subsequently recalled, the navy wanted to develop this "gadget" into a combat weapon immediately:

> So the Navy, as an emergency measure, confiscated, or what do you call it, "requisitioned" a number of yachts and other private boats, small boats, most of them not much more than 100 feet long. I think they were called PC boats. They put sonic gear on them, but they had no way of arguing with a submarine.
>
> Now it happened that we found out that the British had developed a weapon system called the Hedgehog, which works sort of on the principle of an inverted mortar. You fire the barrel with the warhead. They were used on some of our destroyers and were apparently very successful. They could not, however, be used on these small boats because the recoil was enough to break up the wooden structure, sink the boat probably.
>
> We thought we could just substitute a small rocket for these inverted mortars and that worked very well.

An integral part of the Caltech rocketeers' gadget was the launcher, which Captain Shumaker, of the Bureau of Ordnance, dubbed the Mousetrap, because the launcher for the antisubmarine rocket in its firing position really did resemble a mousetrap. The name took among naval officers, sailors, and scientists, who referred to the ammunition—the rocket-propelled antisubmarine bombs—as "Mousetrap ammunition." Lauritsen's group also spent considerable time developing fuses to use on the Mousetrap rocket-propelled projectiles. Further tests of the Mousetrap rocket at Key West, Florida, at the end of April 1942 persuaded the navy to make them standard equipment on submarine chasers and other ships. The Caltech group conducted these tests and then returned to Pasadena, leaving Ralph Smythe behind.

Smythe remained at Key West through the summer, supervising Mousetrap installations and training officers and crews in the use of the new weapon. One of Smythe's jobs was to demonstrate the Mousetrap antisubmarine rocket to visiting navy dignitaries. He took the varied work in stride. He wrote in longhand to Willy Fowler on one occasion,

> Charge off more ammunition to the entertainment budget. We started this week by putting on a show for the Vice Admiral in charge of supply and his staff. Everything went off well with no misfires. He seemed impressed. We had more men available than when Sec'y Knox was here last week and fixed things fancier, including a tent for them to stand under. We have trained five gun crews at the range this week. Every man loads and unloads and we show them what can go wrong. Those suppository primers work O.K. We have had no misfires. . . . Meanwhile 5 mousetraps have arrived so we now have 16 on hand. . . . We finished installing minnie mousetraps [subcaliber ammunition developed for training at sea] on the 450 today. The crew is getting sensitive from being kidded about them and have invented a story about a superexplosive which they are to use. . . . The latest projectors [launchers] look good and we have never had a dud from the service ammunition so it seems to me we ought to start hunting subs when the crews are ready.

In fact, the Mousetrap antisubmarine rocket was hunting subs in domestic waters by autumn of 1942, and it was turned loose on the enemy in the Pacific six months later. Used in conjunction with depth charges, Caltech's antisubmarine rocket was credited with many assists in submarine attacks and was still in use when the war ended.

Increased demand for rocket weapons in combat turned Caltech's rocketry group and the navy into the best of friends. Following a demonstration of the Mousetrap, the navy's commander of the amphibious forces, Admiral Wilson Brown, turned to the group for help with a weapon to be used with amphibious land-ings. As Brown described the scenario, the beach would be bombarded from large ships until the troops got to within a thousand yards of shore. But from there to the beach, the landing troops needed protective cover. "The obvious solution is a small rocket bomb," Charles Lauritsen wrote to Tolman in June 1942, enclosing a drawing of a barrage bomb that his Caltech group was getting ready to test. Development of barrage rockets and launchers for use on support boats and amphibious tractors and trucks proceeded apace, helped greatly by the fact that the rocket motor used in the Mousetrap could also be used, with only minimal changes, in the new rocket weapon. Installed on landing craft and patrol boats, the barrage rockets were first used in the invasion of North Africa in 1943; later they were pressed into service in the storming of the beaches in Sicily and at Salerno. In the Pacific, Caltech-designed rockets figured in more than a score of landing operations, from Guam to Iwo Jima and Okinawa.

When one officer from the Seventh Fleet was asked if the high dispersion of the barrage rocket made it an unsuitable weapon against enemy barges, he replied, "No, because it doesn't make much difference whether we get a direct hit or not. The Japs have got so that as soon as we open up with rockets, they all go overboard. Then we can come up alongside and sink the barge with automatic weapon fire, or chop a hole in the bottom with a hatch."

In spring 1944, the rocket project's assistant director, William Fowler, spent three months in the South Pacific, talking to sailors and marines who were already familiar with rocket weapons. There he learned a lot about what they liked and didn't like about them. At Nouméa, Admiral Halsey and his staff were enthusiastic about the use of aircraft rockets on land. "The results against exposed land targets such as . . . radar installations, and bivouac areas have been spectacular," Fowler wrote back to Lauritsen. But their performance at sea was another story. "The results against enemy shipping and barges have been disappointing," Fowler reported. Put a delay fuse on the aircraft rockets, the pilots told Fowler, who relayed the suggestion back to the group in Pasadena. He also sent back tales of inadequate supplies of ammunition. Halsey, he reported to Lauritsen, had "literally cussed a blue streak because he can't get enough rockets for his operations out here." Fowler urged Lauritsen to use his influence. "Now is the time when rockets can really raise hell."

Caltech's aircraft rockets went into combat action in the Pacific early in 1944. During one daylight shipping strike at Keravia Bay, in the Rabaul region of New Britain Island, two of the five pilots in the first squadron trained in aircraft rocket firing saw their rockets score a hit. Describing his attack run, one of them said, "Coming out of the dive, I sighted on the ship's side and let the rockets go. . . . A second later I pulled the bomb lever. The rockets and bombs smacked the ship like the old one-two in boxing."

Rockets could help to win the war, but they could not end it—Fowler needed no convincing on this point. He felt just as strongly, as did Charles Lauritsen, that it was important for Caltech to get out of the rocket business. Fowler told an interviewer many years later, "We felt that we had done the proper thing in setting up for the Navy a laboratory [the Naval Ordnance Test Station at Inyokern] that could carry on in the rocket developments, and I think in large measure that has paid off." Thus, over Earnest Watson's objections, the closing out of Caltech's rocket program was well along by the end of 1944.

The proximity fuse also chalked up major successes during the war. Commenting in 1944 in a letter to the chief of ordnance on its effectiveness as a battlefield weapon, General George S. Patton noted,

> The new shell with the funny fuse is devastating. The other night we caught a German battalion, which was trying to get across the Sauer River, with a battalion concentration and killed by actual count 702. I think that when all armies get this shell we will have to devise some new method of warfare. I am glad that you all thought of it first.

Charles Lauritsen's decision in December of 1944 to become involved in J. Robert Oppenheimer's atomic bomb project at Los Alamos hastened the transfer of the rocket work to the navy. On December 31, 1945, all of the operations of the Caltech rocket project formally ended.

Fowler remembers that, when the war against Japan ended, in August 1945, he invited Max Mason and Charles Lauritsen and several others over to his house for drinks. "Right then and there the question arose, 'Well, now what are we going to do?' " Many people at Caltech were asking the same question.

FOURTEEN

◆

The DuBridge Era
at Caltech

What stands out in my mind is that at the end of the Millikan era, which coincided with the end of the Second World War—or to put it differently, at the beginning of the Lee DuBridge era—Caltech was taken over by a group of people who had had extensive wartime experience. . . . They all had a sense of the university's relationship to the nation.

—RODMAN W. PAUL, 1982

The first responsibility of the scientist or engineer is to be a *good* scientist or a *good* engineer. . . . It is not the job of the scientist to be primarily a politician, a sociologist, a military leader or a preacher.

—LEE A. DUBRIDGE, 1947

The disposition of Lee and Dr. Millikan are completely and entirely different. Lee is a faculty man and very much needed when he arrived; Dr. Millikan was a trustee man. I think Lee is without guile, and Dr. Millikan was endowed with a lot of it.

—JAMES R. PAGE, 1961

CALTECH'S HISTORY is divided into two distinct eras. The first Caltech era was created by the astronomer George Ellery Hale, the physicist Robert Andrews Millikan, and the chemist Arthur Amos Noyes in the 1920s. Thirty years later, after World War II, the job was done all over again by the physicists Lee Alvin DuBridge and Robert Bacher.

An educational institution in name only during the war, Caltech had a war arsenal that included rockets, proximity fuses (a small device built into the projectile, causing it to explode at a set distance from the target), the Jet Propulsion Laboratory, and $80 million in federal funds for war-related research and development.

The Jet Propulsion Laboratory, as it has been called since 1944, started out as a small group of Caltech graduate students and local amateurs interested in high-altitude sounding rockets, organized more or less loosely in 1936 by the school's Guggenheim Aeronautical Laboratory and named, after it, the GALCIT Rocket Research Project. Spearheaded by Frank Malina, a graduate student in aeronautics at Caltech, the rocketry group included Apollo Smith, Hsue-shen Tsien, and Weld Arnold, all Caltech graduate students, and two local enthusiasts by the names of Edward Forman and John Parsons. Forman was a skilled mechanic. Parsons, a self-taught chemist, had a vast store of information about explosives. Theodore von Kármán, the Hungarian-born engineer and applied scientist and the first director of Caltech's school of aeronautics, offered Malina's group moral support, but no money. So equipment, mostly secondhand, and chemicals were bought with whatever cash the group could scrape together.

Forman and Parsons wanted to shoot off rockets. Malina, backed by von Kármán, insisted on designing a workable motor first. Unlike jet engines, which burn oxygen from the atmosphere, rocket engines carry their own oxygen, and so are capable of soaring beyond the atmosphere. In other words, building a space rocket depended first on solving the rocket engine problem. Armed with master's degrees in mechanical and aeronautical engineering, Malina was also studying problems of rocket propulsion and flight for his Ph.D.—under von Kármán's supervision. All in all, it was an unlikely combination of talents. Indeed, many years later, Malina recalled that the GALCIT rocket group was "looked upon at Caltech as rather odd, to say the least." And possibly as dangerous as well.

Within six months, Malina's group was doing experiments with liquid-fuel rocket motors in an isolated spot in the Arroyo Seco, on the western edge of Pasadena, about three miles above the Rose Bowl. On their last test at the arroyo, on January 16, 1937, the motor ran for forty-four seconds. But after this triumph, and in spite of theoretical studies and testing that continued on the campus for several more years, the rocket project seemed to lose its momentum.

Then it attracted the attention of the Army Air Corps, following a visit by the corps' commanding general, Henry H. Arnold, to the GALCIT laboratories in 1938. A year later, GALCIT received funds from the National Academy of Sciences Committee on Air Corps Research to study the possibility of using rockets to boost the takeoff of large heavy bombers from short runways. But *rocket* wasn't a respectable word in 1939, so the rocket project became the Air Corps' Jet Propulsion Research Project. The project also moved off campus once again, to a new home in the Arroyo Seco, not far from the site of the original rocket motor firing, the home of today's Jet Propulsion Laboratory. In 1940, the Army Air Corps assumed direct control of the rocket project; the solid-propellant rocket engines themselves, called JATOs, for jet-assisted takeoff, were developed, tested, and put in the hands of the navy for use in lifting planes faster from carriers in the Pacific three years later.

Success with booster rockets brought a new contract from the Army Ordnance Department in 1944, this time to develop a series of guided and unguided solid- and liquid-propellant rocket missiles, appropriately named Privates, Corporals, and Sergeants. Both von Kármán, the project's first director, and Malina, its chief engineer, tried to use the army request as a springboard to establish an independent Caltech-owned Jet Propulsion Laboratory. But the school's trustees declined to do so. JPL remained an army research and development laboratory until the late fifties, when it was transferred to NASA and embarked on a new career as the center of America's unmanned space program.

When Robert Millikan, the school's head, officially retired in 1945, after twenty-four years in office, he left behind a scientific institution of the first rank, in search of its future.

The unspoken question hanging over the search committee set up to find a successor to Millikan was this: could a small elite school committed to basic scientific work adapt itself to the changed conditions of the postwar world? Problems ranging from Institute financing and inadequate salaries to new appointments, all brushed aside during the war, haunted the campus.

In physics, in particular, Caltech had in the late 1930s lost ground to other comparable schools. As Bacher tells the story, that Caltech "did as well [then] as it did . . . is a great compliment to the faculty. . . . But especially to the faculty members who learned how to do first-rate work with almost nothing to support it." Charles C. Lauritsen's work in nuclear physics during this period is a classic example of shoestring-and-sealing-wax physics.

In a very real sense, then, the war marked the end of Millikan's era. The new era found its leaders in the war's scientific establishment: Vannevar Bush, James Conant, I. I. Rabi, and Lee DuBridge.

For Millikan, long accustomed to the support of the Rockefeller and Carnegie foundations and other private sources of financial aid, the notion of a government agency's supporting peacetime research was unthinkable. That source of funding would have to be forfeited. But by the war's end, Millikan's reign as Caltech's chief executive officer was already over. He had long since ceased to grasp the magnitude of the federal government's financial involvement with the school's rocket research programs or to deal effectively with the wartime factory that Caltech had become. Even as early as 1941, the day-to-day running of the school was really in the hands of Earnest Watson, dean of the faculty, and de facto chairman of the physics division, and James Page, chairman of the Institute's board of trustees. In 1944, Page orchestrated a palace coup that left Millikan, in Page's words, "definitely and completely out of the administration." Millikan "had deteriorated mentally," Page later recalled, and the other trustees had instructed him to replace Millikan. "This was about the hardest job I ever had to do, and that I was able to do it without blood, guts and feathers all over the lot was due first to Mrs. Millikan's and Clark Millikan's understanding of the case, and to the very good disposition of Dr. Millikan himself," added Page. The school's bylaws had also been changed to provide for a president of the school.

By this time, however, Caltech had really become an extension of Millikan's personality. What chance did a new man have at the job? the Rockefeller Foundation's Warren Weaver asked Page, knowing something about Millikan's style of managing the Institute's affairs. Page replied, "RAM understands it in the front of his head but not in the back, but bit by bit it is being borne in on him. . . . If a new man is pushed around it will be his own fault. After all, Dr. Millikan is 78 years old." On September 15, 1945, Millikan officially retired as administrative head of Caltech, clearing the way for the trustees to find a successor.

After thoroughly canvassing the country, Page tapped the physicist Lee A. DuBridge, who was forty-four, as Caltech's president. Indeed, DuBridge had few serious competitors; perhaps the most remarkable thing about the whole exercise was the near unanimity of all those polled. While the elders of the scientific and scholarly community often listed more than one candidate for the job, DuBridge's name headed everyone's list, from those of the Carnegie Institution's head, Vannevar Bush, and Supreme Court Justice Felix Frankfurter to those of Frank Jewett, president of the National Academy of Sciences, and MIT's president, Karl Compton. Page was more interested in selecting a scientist who knew how to run an institution than in selecting the brightest scientist. "It seems to me too bad to spoil a top hole scientist by making him administrator," he commented to Warren Weaver of the Rockefeller Foundation, who countered, "DuBridge is an excellent experimental physicist, but not a genius. He is, however, a genius at administration. He is calm, wise, and exceedingly adroit at choosing and handling men."

The roots of Lee Alvin DuBridge's diverse talents went back to the Midwest. He was born in Terre Haute, Indiana, in 1901. The son of a YMCA athletics instructor, DuBridge grew up in a succession of cities, excelled in science in high school in Sault Sainte Marie, Michigan, and attended Cornell College in Mount Vernon, Iowa, where he majored in physics. Following graduation, in 1922, he entered the University of Wisconsin on a teaching assistantship. There he mastered the intricacies of atomic structure in Charles Mendenhall's class, which also meant learning scientific German in order to follow the assigned text—Arnold Sommerfeld's 400-page *Atombau und Spektrallinien*. He also took all the standard graduate courses in physics of that period, including ones in heat, thermodynamics, electricity and magnetism, statistical mechanics, and mathematical physics, taught by L. R. Ingersoll, J. R. Roebuck, Max Mason, and Warren Weaver.

Mason's teaching style stuck in DuBridge's memory. "Mason talked rapidly, wrote rapidly on the blackboard, often made mistakes," DuBridge later wrote, recalling his teacher. "He made no apologies for this. In fact, he said, proudly, 'The worse I teach, the more you will learn.' We found this to be true, for we worked long hours over our lecture notes, filling in missing equations, working tough problems, discovering he should have written a minus instead of a plus sign every now and then, etc. But we did learn the subject!"

A quick learner, DuBridge raced through graduate school, earning a master's degree in 1924 and completing his dissertation research on the photoelectric properties of platinum in the fall of 1925. He continued to do photoelectric experiments for another fifteen years. DuBridge's doctoral work involved measuring the emission of electrons from the surface of platinum, using different samples of the metal. By means of the best vacuum techniques then available, DuBridge determined the minimum frequency required to remove an electron from the metal's surface, establishing in the process that electron emission was a genuine physical phenemonon. That September, at the age of twenty-four, DuBridge successfully defended his thesis, mailed it off to the *Physical Review* for publication, and married his college sweetheart, Doris May Koht.

He went on to spend two years at Caltech (1926–28) as a National Research Council fellow, under Robert Millikan's direction, followed by six years in the physics department at Washington University (1928–34), moving up the ranks from assistant to associate professor in 1933. The following year, DuBridge accepted an appointment as professor of physics and chairman of the physics department at the University of Rochester, later becoming dean of the faculty as well.

At Rochester, he promptly hired two theoretical physicists, Fred Seitz and Milton Plesset; when Seitz left for a job elsewhere, DuBridge recruited, on Niels Bohr's advice, a Hitler refugee from Austria, Victor Weisskopf. Bohr, who worked hard in the thirties, in Weisskopf's words, to "sell his Jews to American universities," found in DuBridge, who had already hired other Jewish faculty at Rochester, a receptive administrator. "Bringing Vicky to America was one of my proud achievements," DuBridge later declared. Bohr had no luck when he urged another physics chairman, Robert Millikan, to find a place for the cosmic-ray physicist Bruno Rossi at Caltech.

Along with his other university duties, DuBridge also found time at Rochester in 1935 to embark on a new career as a nuclear physicist. Inspired by the work of the Berkeley physicists Ernest

Lawrence and Don Cooksey, DuBridge arranged for Rochester to build a cyclotron—a machine designed to accelerate charged particles to very high energies, thereby inducing nuclear reactions. "By the fall of 1938," DuBridge later wrote, "we had the equipment in operation, producing protons of energy of about five million electron-volts—later raised to six or seven. In those days this was the highest energy proton beam in existence." Indeed, outside of Lawrence's own cyclotron at Berkeley, only Rochester, Cornell, and Michigan had particle accelerators.

Then came the war. Physicists suddenly found themselves working on practical problems that required immediate solutions. "I wanted to be in on things," DuBridge later recalled, and taking a leave of absence from Rochester in the fall of 1940, he moved his family to Belmont, Massachusetts, and set up shop at MIT, where he organized and directed the work of a facility that was officially named the Radiation Laboratory but was quickly dubbed the Rad Lab. DuBridge's wartime laboratory specialized in the development of radar (an acronym for *ra*dio *d*irection *a*nd *r*anging) equipment in the centimeter-wavelength range. By the war's end, the Rad Lab had grown from an initial group of fifteen physicists to a staff of four thousand and had devised and put into production more than a hundred different types of microwave radar for military use on planes, on ships, and on the ground.

In early 1946, DuBridge returned to the University of Rochester, determined to go back into research and teaching. Speaking from first-hand experience at Columbia University, the physicist I. I. Rabi warned his colleague about what lay in store back home. "You will find it a little difficult at first," he predicted, "to settle down to the circumscribed routine of a university physics department." He spoke the truth, as DuBridge quickly discovered. The freshman lectures were no longer fun to give, and the lecture demonstrations, his pride and joy, didn't always work. Worse still, the cyclotron business had changed dramatically since 1940. He despaired of ever being able to catch up.

All the same, the Caltech trustee James Page's real work came in persuading DuBridge to become an administrator again. But Vannevar Bush, for one, didn't think the prospect of another administrative post was the source of DuBridge's hesitation. DuBridge's reluctance to leave Rochester, Bush told Page, stemmed from the realization "that at the California Institute he would have quite a job on his hands to keep the institution before the proper section of the public in such a way that support would be bound to be forthcoming." It came down to money—who would pay Caltech's bills in the postwar era went to the very core of the problems DuBridge knew he would face in office.

And once raised, the money would pose new problems for academic administrators all over the country in the postwar world. For one thing, to accept federal aid carried its own risks. "No one knows exactly what the shape of future financing will be," the physicist Carl Overhage, another veteran of the Rad Lab, told Page in 1945, "but it seems clear that the essential problem will not be to get the funds, but rather to preserve the full independence and the traditional character of academic institutions." DuBridge, he went on to say, had never compromised these basic principles during his tenure at the MIT Radiation Laboratory. Established by the National Defense Research Committee, the Rad Lab had begun and ended its existence, nevertheless, as "a free civilian agency of scientific investigation."

Max Mason played a major role in persuading DuBridge to come to Caltech. Not long after he returned to Rochester, DuBridge received a telephone call from Mason, his former physics teacher at the University of Wisconsin and now a trustee at Caltech. As DuBridge remembers it, Mason started out the conversation by saying, "Lee, you've GOT to come out to Caltech as President." DuBridge promised to think it over. In fact, what DuBridge told Mason over the telephone ("He said the offer was something no man could ignore and he is not saying he is not interested," Mason informed Page) raised Page's hopes, and he arranged a meeting with DuBridge in New York in March. Later that month, DuBridge visited Pasadena, talked to trustees and old friends, and toured the campus. He thought the invitation over for several weeks longer, spoke countless times to Mason over the telephone, and finally wired Page to say he had "decided to consider favorably" Caltech's offer. On September 1, 1946, Lee A. DuBridge became president of Caltech, at a starting salary of $20,000.

DuBridge was Caltech's first president, and he served in the post for twenty-three years, retiring only to become President Nixon's science adviser in 1969. Millikan had all along refused the title of president, preferring instead to administer the school's affairs through a faculty-trustee executive council, of which he was chairman. This governing system had puzzled many. Somebody once asked the Caltech physicist Richard Tolman, "What is a Chairman of the Executive Council?" Tolman replied, "Well, a Chairman of the Executive Council is just like a president, *only more so.*"

Under DuBridge, it worked differently, more the way it did at other places of higher learning. And, like those of other universities, many of the school's immediate needs in 1946, when DuBridge arrived, touched on money. The University of Rochester, DuBridge told Caltech's trustees, expected to spend $750,000 just to start up its postwar program in physics. Caltech's combined physics, mathematics, and astronomy budget, by comparison, barely exceeded $200,000, even when Robert Bacher came as division chairman several years later. Page had assured Rockefeller officials there was sufficient money on hand to cover expenses for a year and a half through 1947 provided DuBridge didn't try to do everything at once. After that, the crystal ball turned cloudy. "In any event," he told Warren Weaver, "eighteen months before you starve is quite a time."

In 1946, the school's endowment amounted to $17 million; the operating budget was less than $8 million. Caltech had spent none of its own money during the war—the costs of the military projects having been covered by government contracts. As a result, the school had accumulated reserves amounting to more than $1 million, and substantial salary increases went into effect shortly after DuBridge took office.

DuBridge also put the school on a twelve-month salary plan. In those days, the nine-month academic salary was standard practice, meaning that teachers were paid only during the school year, usually September through May. Meanwhile, federal support for peacetime science had begun. Already in 1946, the Office of Naval Research was supporting Charles Lauritsen's nuclear physics group, as well as other campus projects. As these research contracts initially paid only the summer salaries, when class was out, many of Caltech's professors were, in practice, already receiving a twelve-months' salary. DuBridge simply institutionalized the practice for all the scientists. In effect, the government played banker to Caltech's faculty salary budget. Researchers without contracts got the extra two-ninths increase in salary; those with contracts got a raise. Merit increases followed. Overall, DuBridge boosted the wages of the faculty nearly 33 percent, a revealing commentary on the wages of science under Millikan. To be sure, the fraction of research time charged to government contracts has been the subject of intense discussion on both sides ever since.

On the academic side, physics—Caltech's crown jewel—also needed attention. In assessing the situation in 1946, DuBridge wrote, "[Physics] is close to my heart, and is sadly in need of repair. Something needs to be done quickly to save what could be a superb department from collapse." New faculty was essential: the spectroscopist Ira Bowen had resigned to become director of the Mount Wilson and the new Mount Palomar observatories; J. Robert Oppenheimer, the only nuclear theorist, had left to become head of the Institute for Advanced Study, in Princeton; William Houston, a versatile solid-state experimentalist and theorist, had gone to Rice as president, leaving the physics chairman's post vacant as well; and Harry Bateman, the famed mathematical physicist, had died.

There remained three superb experimental physics groups:

Lauritsen's in nuclear physics, Carl Anderson's in cosmic rays, and Jesse DuMond's in gamma rays.

When Robert Bacher arrived in Pasadena in summer 1949 to take up his duties as chairman of the division of physics, mathematics, and astronomy, his "reception," he recalls, "was an August heat wave and the absence of most faculty except those from the Kellogg Laboratory [Lauritsen's group], who never seemed to believe in vacations," which suited Bacher fine since he was himself known for his dedication.

Like Lee DuBridge, Robert Bacher had been a National Research Council fellow at Caltech early in his career. An atomic spectroscopist turned nuclear physicist, he had guided Cornell's nuclear physics laboratory to prominence in the 1930s. DuBridge persuaded him in 1941 to come to MIT's Radiation Laboratory. In 1943, Bacher had gone, like so many others, to Los Alamos, where he initially served as an adviser to Oppenheimer. Before long, he turned all of his attention to the atomic bomb project. Bacher's division, the physicist Hans Bethe recalls, "had to do with physical measurements by which you could decide whether the assembly was satisfactory. . . . There were many, very good experimentalists . . . [in the division]. But some of them were rather difficult to deal with. Somehow Bob dealt with them." When DuBridge joined Caltech, Bacher was at the top of his list as Caltech's new physics head.

But Bacher proved hard to get.

He turned down the chairman's job in 1946, because he had agreed to serve on the new Atomic Energy Commission. He was the only scientist on the five-man commission. DuBridge persevered. By early 1949, Bacher had told President Truman he wanted to return to academic life, had resigned from Cornell as professor, and had made up his mind to join DuBridge in Pasadena.

Bacher rebuilt the physics department, and he did so with a vengeance, starting with high-energy particle physics. Then a new field, particle physics hardly existed at Caltech in 1949, except for the work of Anderson and his students, including Donald Glaser, a later Nobel Prize winner, who used cosmic rays from space as a natural source of high-energy particles to do particle physics. Bacher resolved to build an electron synchrotron, like the one he had started at Cornell, so that the Caltech group could make its own high-energy particles. Indeed, Charles Lauritsen had already hired Robert Langmuir to draw up some designs for such a machine. Robert Walker, a young Cornell experimentalist whom Bacher had met at Los Alamos, was the first addition to Caltech's high-energy staff. Others followed, including Matthew Sands and Alvin Tollestrup. Like Walker, Sands had also worked at Los Alamos, earning there a reputation for being not only a good physicist but also a wizard at designing circuits, an indispensable asset in a high-energy laboratory.

From the beginning, Caltech's experimental research in high-energy physics centered on observing some of the simplest nuclear reactions involving unstable particles in atomic nuclei. "One of the reasons for building the synchrotron," Bacher wrote in 1957, "was to see whether the unstable heavy mesons (K-particles) were produced by photons and, if so, to study this production process." Caltech's machine reached its full energy in the late 1950s, by which time the next generation of high-energy machines had come off the drawing board and were under construction in this country and abroad. This was an early round in the game of leapfrog that builders of high-energy accelerators have played ever since. Another machine wasn't the answer for Caltech. Instead, the school began in the early 1960s the practice of sending its high-energy experimentalists to big machines elsewhere. The Institute closed down its electron synchrotron in 1969, shortly after ground was broken for the national accelerator laboratory—Fermilab, in Batavia, Illinois.

Theoretical physics, always a stepchild under Millikan, finally came into its own during DuBridge's time. Richard Feynman, then at Cornell, was Bacher's first acquisition. "Feynman was a terrible loss," his colleague Bethe once said. "And a loss that we did not expect. After all, he had done his most fundamental work at Cornell." Ithaca's loss was a gain for Pasadena, which now boasted two of the brightest young physicists from the theoretical group at Los Alamos, Robert Christy having taken Oppenheimer's place in 1946.

Christy was born in British Columbia, Canada, and had been educated at the University of British Columbia and the University of California at Berkeley. At Los Alamos, his calculations were so critical to the success of the bomb that, for a while, those working on it called it "the Christy bomb." At Caltech, he became the Institute's tie to modern theoretical physics. He started out as a member of the Kellogg Laboratory. But Christy's influence extended far beyond theoretical problems connected with nuclear physics. Bacher has said, "I never considered thinking about anything that we should initiate in appointments that had anything to do with new fields or theoretical people without first talking to Christy about it."

Physics has always been central to Caltech's undergraduate curriculum. In the early 1960s, the basic physics course required of all freshmen and sophomores underwent major surgery at the hands of Feynman. He spent two years revising the content of the two-year introductory course and another two, in 201 East Bridge, lecturing to the students and untold faculty and staff besides. It was a noble experiment. A stiff dose of quantum mechanics found its way into the new course, organized along the lines of mathematical and conceptual techniques that Feynman had himself invented. The lectures themselves were taped, transcribed, and edited by his colleagues—primarily Robert Leighton and Sands—and subsequently published in three volumes, set off by bright red covers recognizable on sight to physicists around the world. Feynman's course was tough, even for Caltech students. It, too, has passed into history, but not the *Feynman Lectures on Physics*. Several decades later, more than half a million copies in English have been sold; it has been reprinted twenty-seven times, and half a dozen foreign-language editions, including Japanese, Hungarian, and Spanish, are in circulation throughout the world.

Before the inauguration of the 200-inch telescope, in 1949, Caltech had no astronomy department to speak of, although it always had faculty who worked in astronomy. Tolman was interested in the entropy of the universe; Fritz Zwicky, in neutron stars. Tolman was a relativity expert who worked with Edwin Hubble, the Mount Wilson astronomer; Zwicky had developed his ideas largely in isolation, along with a reputation as a "pretty tough character." But it was Jesse Greenstein, a distinguished astronomer in his own right, who came to Caltech in 1948, who built the field from the ground up into the formidable enterprise it became during the DuBridge years. As with high-energy physics, the theoretical and the experimental work were built up together. H. P. Robertson, who promoted with great success the application of general relativity to astronomy, joined the faculty, as did Maarten Schmidt, the astronomer who would discover the large redshift in the light from objects that came to be called quasi-stellar objects, or quasars.

When DuBridge arrived in 1946, Linus Pauling was chairman of the division of chemistry, and George Beadle had recently succeeded Thomas Hunt Morgan as head of the biology division. Beadle's work had cemented the idea that genes control enzymes, the chemical stuff of life. Together, Pauling and Beadle established a new program in chemical biology. Pauling went on to help found the field of modern molecular biology. In the humanities, DuBridge recruited Hallett Smith, an outstanding Elizabethan scholar as its new head; in geology, Robert Sharp. I. I. Rabi once remarked of DuBridge, "He believed in his people and what they could do." Geology is a case in point.

In 1950, the division took a chance on a new field, geochemistry. Geologists elsewhere scoffed. "I would go to national geological meetings," Sharp recalls, "and geologists would come up and hiss in my face, 'How's the department of geochemistry at Caltech?' " "Just be patient and give us time," Sharp would reply. Within the decade, diverse problems, ranging from the temperature of the oceans in prehistoric times to the age of trees swallowed up by advancing glaciers, had become the staple food of

Caltech's pioneer geochemists. They had created a truly flourishing field, just as Sharp had predicted. But there is something else worth noting about geochemistry's development at the Institute. It was a new discipline that attracted traditional geologists as well. On the campus, nobody called them geochemists, Sharp once said, although those on the outside "would have, because of the nature of the work they were doing." Not for the first time in Caltech's history, its scientists had built an independent discipline that cut across conventional boundaries. Aeronautics, applied mathematics, and seismology offer similar case histories during DuBridge's administration.

In his ability to explain science to the public, presidents, and members of Congress and to defend the principle of academic freedom during the McCarthy period, DuBridge had few peers. When he spoke, he expressed the aspirations not only of university scientists but of the academic community as a whole. He testified before Congress in 1945 on behalf of a federally supported program of research in pure science, and kept hammering away on this theme, while Congress considered several bills. Speaking at DuBridge's inauguration in 1946, Millikan had reminded his audience of the long historical precedent of "locally tax-supported institutions [of higher education], though up to the present," he emphasized "the grave menace of federal control has been avoided." For DuBridge, on the other hand, the matter of freedom of scientific thought and government support of science went hand in hand. Indeed, from his point of view, freedom of science could be guaranteed only if scientists were not dependent "for its support *solely* on military agencies, or solely on any agency whose *primary* interest is in applied science." Given the expanded role science seemed destined to play in national affairs, including defense, DuBridge argued in 1949, the federal government had a responsibility to support it. When Public Law 501—the act establishing the National Science Foundation—went into effect, in 1950, President Truman appointed DuBridge to the National Science Board, the policy-making branch of the new foundation. He served as chairman of many committees and boards, including the Science Advisory Committee of the Office of Defense Mobilization, established in the early 1950s. In 1957, following the Soviets' launch of an artificial satellite named Sputnik, the Office of Defense Mobilization became the President's Science Advisory Committee (PSAC) and reported directly to him.

During the 1950s, DuBridge had good reason to ponder the question, What are the obligations of a free academic institution when the issue is national defense? In 1951, the armed services asked Caltech to conduct a study of the defense of Western Europe. The project, which became known as Project Vista, dealt largely with tactical military weapons, including the use of nuclear weapons on the battlefield. DuBridge reluctantly agreed to Caltech's sponsorship of the classified project, portions of which were declassified only in 1980.

Caltech's Jet Propulsion Laboratory posed another set of problems. In response to the Soviet detonation of an atomic bomb in 1949, and to the Korean War the following year, the laboratory became a largely secret installation, its engineers bent on converting missiles, like the Corporal, into tactical surface-to-surface weapons. Like it or not, JPL had become an integral part of the country's stepped-up military program—and its relationship with the army and Caltech had grown increasingly controversial. Reacting to criticism by some of the faculty and trustees, DuBridge vigorously defended the school's management of JPL, pointing out that weapons development was "not the fault of Caltech or of the military services but the fault of one Joseph Stalin." Still, the price of cooperation came high. In 1953, the army tried to put JPL under military control. The attempt failed because the laboratory's director, Louis Dunn, and DuBridge said Caltech would end its military connections. The army conceded the point, partly because it enjoyed the prestige of having Caltech on its payroll. Financially, the school had in fact become more dependent on the negotiated overhead fee the army paid it to manage the laboratory. JPL, for its part, seemed determined to go its own way, regardless of the other players.

DuBridge also testified on behalf of Robert Oppenheimer,

over the strenuous objections of some of Caltech's own trustees, when Oppenheimer was denied security clearance in 1954. He was prepared to resign, rather than retract his public words of support. He defended Pauling's right to speak out on unpopular causes against a divided board of trustees, and a divided faculty, and in the process put his own job on the line. In its cover story on Caltech's president in 1955, *Time* magazine dubbed him "Senior Statesman of Science."

Millikan, too, had once appeared on the cover of *Time*. The point is, although Millikan and DuBridge may have belonged to different generations, each symbolized the scientific establishment of his own time. The power structure Hale, Millikan, and Noyes planned, built, and ran for two decades in America is a familiar theme in books and articles recounting the rise of science in America in the first part of this century. Less well known are the ways in which the size, shape, and tempo of science changed after the war. Those changes, powered by the likes of DuBridge and Bacher, led to altogether different enterprises, in an altogether different world.

SELECT BIBLIOGRAPHY

♦

Adams, Walter S. "Early Days at Mount Wilson." *Publications of the Astronomical Society of America* 59 (1947): 213–31.
———. "George Ellery Hale, 1868–1938." *National Academy of Sciences Biographical Memoirs* 21 (1940): 181–241.
Allen, Garland E. *Thomas Hunt Morgan: The Man and His Science.* Princeton, 1978.
Anderson, John A. "The Astrophysical Observatory of the California Institute of Technology." *Journal of the Royal Astronomical Society of Canada* 36 (1942): 177–200.
Badash, Lawerence. "The Age-of-the-Earth Debate." *Scientific American*, Aug. 1989, pp. 90–96.
———. "Nuclear Physics in Rutherford's Laboratory before the Discovery of the Neutron." *American Journal of Physics* 51 (1983): 884–89.
Barnes, C. A., D. D. Clayton, and D. N. Schramm, eds. *Essays in Nuclear Astrophysics: Presented to William A. Fowler, on the Occasion of his Seventieth Birthday.* Cambridge, Eng., 1982.
Baxter, James P. *Scientists against Time.* Boston, 1946.
Bethe, Hans A. "Energy Production in Stars." *Physical Review* 55 (1939): 434–56.
———. "Energy Production in Stars." *Science* 161 (1968): 541–47.
Bethe, Hans A., and C. L. Critchfield. "The Formation of Deuterons by Proton Combination." *Physical Review* 54 (1938): 248–54.
Bilstein, Roger E. *Flight in America, 1900–1983.* Baltimore, 1984.
Burchard, John E. *Rockets, Guns and Targets.* Boston, 1948.
Burns, James MacGregor. *The Workshop of Democracy.* Vol. 2 of *The American Experiment.* New York, 1986.
Carlson, Elof Axel. *Genes, Radiation, and Society: The Life and Work of H. J. Muller.* Ithaca, 1981.
Christman, Albert E. *Sailors, Scientists, and Rockets.* Vol. 1 of *History of the Naval Weapons Center, China Lake, Calif.* Washington, D.C., 1971.

Clark, David L. *The Aerospace Industry as the Primary Factor in the Industrial Development of Southern California: The Instability of the Aerospace Industry, and the Effects of the Region's Dependence on it.* Los Angeles Community Analysis Bureau. The Economic Development of Southern California, 1920–1976. Vol. 1. Edited by John Bruce. Los Angeles, 1976.

Davis, Nuel Pharr. *Lawrence and Oppenheimer.* New York, 1968.

DuBridge, Lee A., and Paul S. Epstein. "Robert A. Millikan, 1868–1953." *National Academy of Sciences Biographical Memoirs* 33 (1959): 241–82.

Elliot, David C. "Project Vista and Nuclear Weapons in Europe." *International Security* 11 (1986): 163–83.

Erdélyi, Arthur E. "Harry Bateman 1882–1946." *Obituary Notices of the Fellows of the Royal Society* 5 (1948): 591–618.

Fehrenbach, T. R. *F.D.R.'s Undeclared War, 1939 to 1941.* New York, 1967.

Fowler, William A. "Charles Christian Lauritsen, 1892–1968." *National Academy of Sciences Biographical Memoirs* 46 (1975): 221–39.

Fuller, Mardi Bettes. "Theodosius Dobzhansky Papers." In *Mendel Newsletter*, June 1980, pp. 1–7. Issued by the American Philosophical Society Library, Philadelphia. Mimeo.

Geiger, Roger L. *To Advance Knowledge: The Growth of American Research Universities, 1900–1940.* New York, 1986.

Gerard-Gough, J. D., and Albert E. Christman. *The Grand Experiment at Inyokern.* Vol. 2 of *History of the Naval Weapons Center, China Lake, Calif.* Washington, D.C., 1978.

Greenberg, John L., and Judith R. Goodstein. "Theodore von Kármán and Applied Mathematics in America." *Science* 222 (1983): 1300–304.

Hafstad, L. R., and M. A. Tuve. "Artificial Radioactivity Using Carbon Targets." *Physical Review* 45 (1934): 902–3.

———. "Induced Radioactivity Using Carbon Targets." *Physical Review* 47 (1935): 506.

Hall, R. N., and W. A. Fowler. "The Cross Section for the Radiative Capture of Protons by C^{12} near 100 Kev." *Physical Review* 77 (1950): 197–204.

Hallion, Richard P. *Legacy of Flight: The Guggenheim Contribution to American Aviation.* Seattle, 1977.

Hanle, Paul A. *Bringing Aerodynamics to America.* Cambridge, Mass., 1982.

Heilbron, John L., Robert W. Seidel, and Bruce R. Wheaton. *Lawrence and His Laboratory: Nuclear Science at Berkeley, 1931–1961.* Berkeley, 1981.

Hill, Laurance L. *Six Collegiate Decades, the Growth of Higher Education in Southern California.* Los Angeles, 1929.

Hirsch, Richard. "The Riddle of the Gaseous Nebulae." *Isis* 70 (1979): 197–212.

Holbrow, Charles H. "The Giant Cancer Tube and the Kellogg Radiation Laboratory." *Physics Today* 34 (1981): 6–13.

Jonas, Gerald. *The Circuit Riders: Rockefeller Money and the Rise of Modern Science.* New York, 1989.

Kargon, Robert H. "The Conservative Mode: Robert A. Millikan and the Twentieth-Century Revolution in Physics." *Isis* 68 (1977): 509–26.

———. *The Rise of Robert Millikan: Portrait of a Life in American Science.* Ithaca, 1982.

———. "Temple to Science: Cooperative Research and the Birth of the California Institute of Technology." *Historical Studies in the Physical Sciences* 8 (1977): 3–31.

Kármán, Theodore von, with Lee Edson. *The Wind and Beyond: Theodore von Kármán, Pioneer in Aviation and Pathfinder in Space.* Boston, 1967.

Kevles, Daniel J. *In the Name of Eugenics: Genetics and the Uses of Human Heredity.* New York, 1985.

———. *The Physicists: The History of a Scientific Community in Modern America.* New York, 1978.

Kohler, Robert E. "A Policy for the Advancement of Science: The Rockefeller Foundation, 1924–29." *Minerva* 16 (1978): 480–515.

Koppes, Clayton R. *JPL and the American Space Program.* New Haven, 1982.

Lauritsen, Thomas. "Kellogg Laboratory: The Early Years." *Engineering and Science* 32 (1969): 4–7.

McDougall, Walter A. . . . *The Heavens and the Earth.* New York, 1985.

McWilliams, Carey. *Southern California Country.* New York, 1946.

Mayr, Ernst. *The Growth of Biological Thought.* Cambridge, Mass., 1982.

Millikan, Robert A. "Albert Abraham Michelson, 1852–1931." *National Academy of Sciences Biographical Memoirs* 19 (1938): 121–47.

———. *The Autobiography of Robert A. Millikan.* New York, 1950.

282   SELECT BIBLIOGRAPHY

Offner, Arnold A., ed. *America and the Origins of World War II, 1933–1941.* Boston, 1971.

Paradowski, Robert J. "The Structural Chemistry of Linus Pauling." Ph.D. diss., University of Wisconsin, 1972.

Pauling, Linus. "Arthur Amos Noyes, 1866–1936." *National Academy of Sciences Biographical Memoirs* 31 (1958): 322–46.

Raymond, Arthur E. "Who? Me?: Autobiography of Arthur E. Raymond." Pasadena, Calif., 1974. Mimeo.

Reid, Hiram Alvin. *History of Pasadena.* Pasadena, 1895.

Reingold, Nathan. "The Case of the Disappearing Laboratory," *American Quarterly* 29 (1977): 79–101.

Rutherford, Ernest. *Radioactivity.* Cambridge, Eng., 1904.

Sandler, Iris, and Laurence Sandler. "A Conceptual Ambiguity That Contributed to the Neglect of Mendel's Paper." *History and Philosophy of the Life Sciences* 7 (1985): 3–70.

Scheid, Ann. *Pasadena: Crown of the Valley.* Northridge, Calif., 1986.

Sears, William R., and Mabel R. Sears. "The Kármán Years at GALCIT." *Annual Review of Fluid Mechanics* 11 (1979): 1–10.

"Seat of Science." *Fortune,* July 1932, p. 18.

Seidel, Robert W. "Physics Research in California: The Rise of a Leading Sector in American Physics." Ph.D. diss., UC Berkeley, 1978.

Seims, Charles. *Mount Lowe, The Railway in the Clouds.* San Marino, Calif., 1976.

Servos, John W. "The Industrial Relations of Science: Chemical Engineering at MIT, 1900–1939." *Isis* 71 (1980): 531–49.

———. *Physical Chemistry from Ostwald to Pauling: the Making of a Science in America.* Princeton, 1990.

Stadtman, Verne A. *The University of California, 1868–1968.* New York, 1970.

Starr, Kevin. *Inventing the Dream: California through the Progressive Era.* New York, 1985.

Stone, Alice, and Judith R. Goodstein, "Windows Back of a Dream." In *Caltech 1910–1950: An Urban Architecture for Southern California,* edited by Jay Belloli, pp. 7–14. Pasadena, 1983.

Stuewer, Roger H., ed. *Nuclear Physics in Retrospect: Proceedings of a Symposium on the 1930's.* Minneapolis, 1979.

Taylor, Geoffrey. "Aeronautics before 1919." *Nature* 233 (1971): 527–29.

U.S. Department of the Interior, Bureau of Education. *Art and Industry:*

Education in the Industrial and Fine Arts in the United States. Compiled by I. E. Clarke. Part 3. Washington, D.C., 1897.

Voegtlin, Carl. "John Jacob Abel, 1857–1938." *Journal of Pharmacology and Experimental Therapeutics* 67 (1939): 373–406.

Weiner, Charles, ed. *Exploring the History of Nuclear Physics.* Proceedings of the American Institute of Physics-American Academy of Arts and Sciences, vol. 7. New York, 1972.

Weinstein, Alexander. "Wilson's Ostrich Egg and Morgan's Omelet." *Quarterly Review of Biology* 55 (1980): 43–55.

Wise, George. *Willis R. Whitney, General Electric, and the Origins of U.S. Industrial Research.* New York, 1985.

Wright, Helen. *Palomar.* New York, 1952.

———. *Explorer of the Universe: A Biography of George Ellery Hale.* New York, 1966.

SELECT BIBLIOGRAPHY

NOTES

The following abbreviations are used: JAA, John A. Anderson Papers; ECB, Edward C. Barrett Papers; HB, Harry Bateman Papers; WAF, William A. Fowler Papers; BG, Beno Gutenberg Papers; GEH, George Ellery Hale Papers; FJM, Frank J. Malina Papers; CBM, Clark Blanchard Millikan Papers; RAM, Robert Andrews Millikan Collection; THM, Thomas Hunt Morgan Papers; AAN, Arthur Amos Noyes Papers; JABS, James A. B. Scherer Papers; AHS, Alfred H. Sturtevant Papers; AGT, Amos Gager Throop Collection; RCT, Richard Chace Tolman Papers; TVK, Theodore von Kármán Collection; HOW, Harry O. Wood Papers—all in the Institute Archives, California Institute of Technology; GEB, General Education Board; RFA, Rockefeller Foundation Archives—all in the Rockefeller Archive Center, North Tarrytown, New York; and UCB, President's Files, University Archives, Bancroft Library, UC Berkeley.

CHAPTER I   From the Orange Groves to "Camp Throop"

Page 23   Headquote: Norman Bridge, "Southern California as a Summer Resort," *Land of Sunshine* 10 (1899):354.

Page 25   ("Sinque of corruption"): Quoted in Shelley Erwin and Carol H. Bugé, eds., *The Amos Gager Throop Collection: A Guide to the Papers in the Archives of the California Institute of Technology and the Chicago Historical Society* (Pasadena, 1990), p. 8.

Page 27   ("Planted potatoes, cleaned a water pipe . . ."): Amos Gager Throop, diary entry, Sept. 1, 1891, AGT, Box 3.

Page 27   ("has commenced this morning . . ."): Throop University, *Minutes of the Executive Committee, 1891–1895*, p. 23, Institute Archives, Historical Files, Box 1.14.

Page 27   ("suspend all action . . ."): Ibid., p. 42.

Page 28   ("a higher appreciation . . ."): G. S. Mills, quoted in *First Annual Catalogue of Throop University*, 1892–93, p. 8.

Page 29   ("talk Throop among the players . . ."): Throop Polytechnic Institute Minute Book B, 1899, Office of the President, Records of the Board of Trustees, California Institute of Technology, Pasadena.

Page 29   ("the larger classes visited . . ."): Cornelius B. Bradley, "Memorandum Concerning Throop Polytechnic Institute," Jan. 5, 1900, UCB, CU-5, Box 36: 64.

Page 30   ("concentrate their entire attention . . ."): G. E. Hale to G. V. Wendell, Jan. 13, 1908, GEH, Box, 43.

Page 30   ("the educational atmosphere . . ."): H. E. Clifford to Hale, Oct. 17, 1907, GEH, Box 11.

Page 30   ("Under the old regime . . ."): Hale to Wendell, Feb. 6, 1908, GEH, Box 43.

Page 31   ("We have no women at all"): J. A. B. Scherer to A. Fleming, Sept. 27, 1910, JABS, Box 1.3.

Page 31   ("I was much pleased . . ."): Hale to Scherer, Sept. 30, 1910, JABS, Box, 2.5.

Page 31   ("California institute of technology"): Assembly Bill 902, Jan. 31, 1911, copy in JABS, Box 3.7.

Page 32   ("Throop stock has gone up 1000%"): Scherer to Hale, Feb. 7, 1911, GEH, Box 36.

Page 32   ("It is certain . . ."): D. S. Jordan to Scherer, Feb. 26, 1911, JABS, Box 3.2.

Page 33   ("a high-grade institute . . ."): Hale to Scherer, May 9, 1908, GEH, Box 36.

Page 33   ("patriotic preparedness"): "The President's Ninth–Tenth Annual Report," *Throop College Bulletin* 28 (1919): 13.

Page 33   ("a new baptism of patriotism"): News clipping, Feb. 10, 1916, Institute Archives, scrapbook collection.

Page 33   ("There are over 1000 . . ."): Scherer to Hale, July 26, 1916, GEH, Box 36.

Page 33   ("We granted their petition"): Scherer to Hale, May 24, 1916, ibid.

Page 34 ("We were being keyed up . . ."): Quoted in "Harvey House," interview by Ruth Powell, 1982, pp. 15–16, Institute Archives.

Page 34 ("not well enough informed . . ."): F. Thomas to Hale, Oct. 2, 1917, GEH, Box 76.

Page 34 ("He has 'the goods,' . . ."): E. C. Barrett to Scherer, Nov. 2, 1917, ECB, Box 2.8.

Page 34 ("the call of Christianity"): James A. B. Scherer, "The Moral Equivalent of War," *Throop College Bulletin* 24 (Jan. 1915): 12.

Page 34 ("The school is broadly Christian"): Scherer, "Introductory," *Throop College Bulletin* 25 (Jan. 1916): 25.

Page 34 ("America has become . . ."): Scherer to W. H. Hays, June 16, 1919, JABS, Box 1.12.

Page 34 ("a preparation for life . . ."): "Marks of Progress," *Throop College Bulletin* 25 (April 1916): 16.

Page 35 ("I am wondering whether . . ."): Scherer to General H. P. McCain, Sept. 28, 1916, JABS, Box 1.9.

Page 35 ("We would get out there . . ."): Quoted in "Earl Mendenhall," interview by Ruth Powell, 1981, p. 11, Institute Archives.

Page 35 ("In spite of . . ."): Scherer to Hale, May 9, 1917, GEH, Box 36.

Page 36 ("that the Adjutant General . . ."): Scherer to Hale, May 10, 1919, ibid.

Page 36 ("discriminatory remarks . . ."): News clipping, June 26, 1918, JABS, Box 3.10.

Page 36 ("German peace"): "An Open Letter to Secretary Baker," *New York Times*, June 26, 1918, copy in JABS, Box 1.11.

Page 36 ("The students are rolling in . . ."): Scherer to Fleming, Sept. 27, 1918, JABS, Box 1.11.

Page 37 ("Soldiers of the United States Army . . ."): Quoted in "Dear Crusader," Oct. 15, 1918, ibid.

Page 37 ("We have had . . ."): Scherer, "Report," Jan. 7, 1919, JABS, Box 1.12.

Page 38 ("that military training . . ."): Ibid.

Page 38 ("I believe there may be . . ."): Edward C. Barrett, "Memorandum for President Scherer," Dec. 5, 1918, ECB, Box 2.9.

Page 38 ("properly prepared students"): Circular letter, Dec. 14, 1918, JABS, Box 1.11.

Page 39 ("We have 230 students . . ."): Scherer to Hale, March 21, 1919, GEH, Box 36; statistics drawn from Barrett to Fleming, March 19, 1919, ECB, Box 1.6 and Barrett to Hale, Nov. 25, 1919, GEH, Box 76.

Page 39 ("an endowment fund . . ."): Barrett to Scherer, Dec. 20, 1916, copy in GEH, ibid.

Page 39 ("to remove the name Throop . . ."): A. A. Noyes to Scherer, Dec. 5, 1919, JABS, Box 2.14.

Page 39 ("It is merely . . ."): George Ellery Hale, "Suggested Notes for Part of Campaign Literature," Nov. 1919, copy in JABS, Box 2.7.

Page 40 ("Hooray for Caltech . . ."): Quoted in "Harvey House," p. 39.

CHAPTER 2 Preamble to a Technical School

Page 41 Headquote: Arthur Amos Noyes to George Ellery Hale, Oct. 15, 1907, GEH, Box 31.

Page 41 ("make no small plans"): Quoted in F. H. Seares, "George Ellery Hale: The Scientist Afield," *Isis* 30 (May 1939): 244.

Page 42 ("helped greatly to arouse . . ."): George Ellery Hale, "Biographical Notes," Feb. 8, 1933, p. 1, GEH, Box 92.

Page 42 ("I was born an experimentalist . . ."): Ibid.

Page 44 ("I was thus bound . . ."): Quoted in Freeman Dyson, "Astronomy in a Private Sphere," *American Scholar* 53 (Spring 1984): 171.

Page 45 ("Rich people who move . . ."): Quoted in Carey McWilliams, *Southern California Country* (New York, 1946), p. 327.

Page 45 ("During those years . . ."): Norman Bridge, *The Marching Years* (New York, 1920), p. 160.

Page 46 ("pointed out how . . ."): Hale, "Biographical Notes," p. 28.

Page 47 ("specialized technical courses . . ."): Quoted in James A. B. Scherer, "The Throop Idea," installation address, Nov. 19, 1908, in *Throop Institute Bulletin* 40 (Dec. 1908): 23.

Page 47 ("Under such conditions . . ."): Ibid.

Page 48 ("We could and should do more"): W. A. Edwards, "Report of President of the Institute," in *Throop Institute Bulletin* 29 (Oct. 1905): 5.

Page 48 ("in a mediocre way . . ."): Hale to J. A. B. Scherer, May 9, 1908, GEH, Box 36.

Page 49 ("educate men broadly . . ."): Ibid.

Page 49 ("If science is to be regarded . . ."): George Ellery Hale, "A Plea for the Imaginative Element in Technical Education," *Technology Review* 9 (Oct. 1907): 478.

Page 49 ("to build up . . ."): "General Statement," *Throop Institute Bulletin* 37 (June 1907): 6.

Page 51 ("working out the educational scheme"): Hale to A. A. Noyes, Jan. 2, 1909, GEH, Box 31.

Page 51 ("Some of the things . . ."): Hale to Evelina Hale, April 11, 1913, GEH, Box 80.

Page 51 ("I don't know . . ."): Hale to Evelina Hale, April 18, 1913, ibid.

Page 52 ("a technical school . . ."): Hale to Noyes, Jan. 2, 1909, GEH, Box 31.

Page 52 ("principles rather than . . ."): Arthur Amos Noyes, "What Is an Engineer?" address to students at Throop, Feb. 6, 1914, in *Throop College Bulletin* 23 (April 1914): 12.

Page 52 ("you would be only . . ."): Arthur Amos Noyes, "Talk to First-Year Students," Dec. 5, 1906, in *Technology Review* 9 (1907): 6.

Page 53 ("problems . . . that develop . . ."): Arthur Amos Noyes, "A Talk on Teaching, Given at a Conference of Members of the Instructing Staff . . . on March 20, 1908," privately printed, in A. A. Noyes, *Collected Papers*, vol. 3, 1907–1910, Institute Archives.

Page 53 ("give tone to the department"): Noyes to Hale, Dec. 17, 1908, GEH, Box 31.

Page 53 ("research spirit"): Hale to Noyes, Dec. 26, 1908, ibid.

Page 54 ("Unknown and still undeveloped"): Hale to Noyes, Jan. 2, 1909, ibid.

Page 55 ("some alert . . ."): Scherer to Noyes, Feb. 15, 1912, JABS, Box 1.5.

Page 55 ("This is a very . . ."): Hale to Scherer, May 9, 1908, GEH, Box 36.

Page 56 ("Noyes would also give . . ."): Hale to Scherer, April 28, 1913, JABS, Box 2.5.

Page 56 ("Had full charge . . ."): Arthur Amos Noyes, "Life Sketch of Arthur A. Noyes," n.d., AAN, Box 1.

Page 57 ("while the country"): A. Fleming to Scherer, Feb. 27, 1914, JABS, Box 1.25.

Page 58 ("avoid the danger . . ."): Scherer to Fleming, May 27, 1915, JABS, Box 1.8.

Page 58 ("the Professors at Throop . . ."): Fleming to Scherer, May 26, 1915 and Scherer to Fleming, June 1, 1915, JABS, Box 1.25.

Page 59 ("cemetery appearance"): Fleming to Scherer, Nov. 5, 1915, ibid.

Page 59 ("Can't tell . . ."): Hale to Evelina Hale, May 5, 1915, GEH, Box 80.

Page 59 ("Throop's superlative opportunity"): Scherer to Fleming, May 5, 1915, JABS, Box 1.8.

Page 59 ("Boston episode"), ("The acceptance of Noyes . . ."), and ("Personally, I feel . . ."): Scherer to Fleming, May 14, 1915, ibid.

Page 60 ("The plan I presented . . ."): Hale to Noyes, May 19, 1915, GEH, Box 31.

Page 60 ("I can see . . ."): Noyes to Scherer, May 21, 1915, copy in GEH, Box 31.

Page 61 ("I do consider it . . ."): Hale to Noyes, May 27, 1915, ibid.

Page 61 ("There are in America . . ."): Arthur Amos Noyes, "Scientific Research in America," report of an address to students at Throop assembly, Dec. 1915, in *Throop College Bulletin* 69 (Oct. 1915): 14.

Page 62 ("industrial research . . ."): Ibid.

Page 62 ("I most enthusiastically . . ."): Hale to Noyes, June 14, 1915, GEH, Box 31.

Page 63 ("who this man might be"): Scherer to Fleming, Dec. 4, 1915, JABS, Box 1.8.

Page 63 ("I have become . . ."): Noyes to Hale, Jan. 6, 1916, GEH, Box 31.

CHAPTER 3 The Birth of Caltech

Page 64 Headquote: George Ellery Hale to James A. B. Scherer, April 26, 1916, JABS, Box 2.6.

Page 64 ("We never appreciated . . ."): Quoted in Mark Sullivan, *Our Times: The United States, 1900–1925*, 5 vols. (New York, 1933), 5:32.

Page 65 ("the first phase . . ."): Ibid., p. 35.

Page 65 ("This country should prepare . . ."): Quoted ibid., p. 228.
Page 66 ("that the entire field . . ."): Hale to W. H. Welch, Feb. 26, 1917, GEH, Box 43.
Page 66 ("I must be present . . ."): Hale to Scherer, Oct. 5, 1916, GEH, Box 36.
Page 66 ("I really believe . . ."): Hale to Scherer, June 7, 1916, JABS, Box 2.6.
Page 67 ("told him so with vigor"): Robert A. Millikan, *The Autobiography of Robert A. Millikan* (New York, 1950), p. 124.
Page 67 ("He finds that he . . ."): R. A. Millikan to Greta Millikan, April 19, 1916, RAM, Box 49.3.
Page 68 ("We must not prepare . . ."): Quoted in Daniel J. Kevles, *The Physicists* (New York, 1978), p. 113.
Page 68 ("to go ahead . . ."): Hale to Evelina Hale, April 26, 1916, GEH, Box 80.
Page 68 ("Now my plan is . . ."): Hale to Scherer, May 30, 1916, JABS, Box 2.6.
Page 69 ("the average American"): Hale to W. R. Whitney, July 4, 1916, GEH, Box 43.
Page 69 ("hardening of the arteries"): Hale to Whitney, June 19, 1918, ibid.
Page 69 ("The spirit of national service . . ."): Hale to Scherer, May 30, 1916, JABS, Box 2.6.
Page 69 ("As long as . . ."): Hale to Scherer, June 30, 1916, JABS, Box 2.6.
Page 69 ("a national service of the highest kind"): Ibid.
Page 70 ("If I could say . . ."): Hale to Scherer, May 30, 1916, JABS, Box 2.6.
Page 70 ("This is great business"): Scherer to Hale, June 6, 1916, GEH, Box 76.
Page 70 ("Throop has been given . . ."): Hale to Scherer, June 30, 1916, JABS, Box 2.6.
Page 71 ("Throop College . . ."): *New York Times*, July 29, 1916.
Page 72 ("I haven't broached . . ."): Hale to Scherer, May 30, 1916, JABS, Box 2.6.
Page 72 ("He is the most *restless* . . ."): Millikan to Greta Millikan, April 19, 1916, RAM, Box 49.3.
Page 73 ("much more expensive . . ."): Hale to Scherer, July 19, 1916, JABS, Box 2.6.
Page 73 ("I think I can arrange . . ."): Millikan to Hale, Oct. 6, 1916, GEH, Box 76.
Page 73 ("I dread telling them . . ."): Millikan to Greta Millikan, Dec. 30, 1916, RAM, Box 49.3.
Page 73 ("If the science men . . ."): Millikan to Greta Millikan, April 1, 1917, RAM, Box 49.4.
Page 74 ("just before you . . ."): Millikan to Hale, Aug. 5, 1919, GEH, Box 29.
Page 74 ("paved the way . . ."): George Ellery Hale, "Biographical Notes," Feb. 8, 1933, GEH, Box 92.

CHAPTER 4   "A Trio of Men"

Page 76 Headquote: Charles D. Walcott, quoted in Edwin Bidwell Wilson to George Ellery Hale, Jan. 20, 1925, GEH, Box 44. The full text of Walcott's remarks are reprinted in Robert H. Kargon, ed., *The Maturing of American Science* (Washington, D.C., 1974), pp. 49–56.
Page 77 ("There are many who believe"): H. E. Howe to G. Dunn, April 30, 1923, copy in RAM, Box 7.5.
Page 77 ("exuberance") and ("fanaticism"): E. B. Wilson to G. E. Hale, Jan. 20, 1925, GEH, Box 44.
Page 77 ("buttonholes every man . . ."): Hale to J. A. B. Scherer, Feb. 5, 1919, JABS, Box 2.7.
Page 78 ("research institution to deal with . . ."): Quoted in Nathan Reingold, "The Case of the Disappearing Laboratory," *American Quarterly* 29 (Spring 1977): 79; G. E. Vincent to R. A. Millikan, Feb. 5, 1918; RAM, Box 7.13.
Page 78 ("I have hardly known . . ."): Millikan to Greta Millikan, Feb. 16, 1918, RAM, Box 49.7a.
Page 78 ("I have become . . ."): Greta Millikan to Millikan, Feb. 14, 1918, RAM, Box 51.4a.
Page 78 ("I can see . . ."): Greta Millikan to Millikan, Feb. 16, 1918, ibid.
Page 78 ("an institution which is free . . ."): Millikan to Vincent, Feb. 18, 1918, RAM, Box 7.13.
Page 79 ("The founding of a new institute . . ."): Ibid.
Page 80 ("I haven't done . . ."): Millikan to Greta Millikan, March 16, 1919, RAM, Box 50.19.

Page 80 ("planning the future . . ."): Millikan to Greta Millikan, Feb. 22, 1918, RAM, Box 49.7a.

Page 80 ("I shall accordingly propose . . ."): G. E. Hale to A. Fleming, March 4, 1918, GEH, Box 16.

Page 81 ("the Foundation was opposed . . ."): Hale to Scherer, March 14, 1918, JABS, Box 2.6.

Page 82 ("it was virtually a contract"): Scherer to Hale, Feb. 22, 1918, JABS, Box 1.11.

Page 82 ("Before long I shall be . . ."): Ibid.

Page 83 ("Rockefeller scheme"): Hale to Millikan, June 21, 1918; Hale to A. A. Noyes, July 8, 1918, all in GEH, Box 29.

Page 83 ("profoundly interested"): Scherer to Hale, April 2, 1918, GEH, Box 36.

Page 83 ("We wished to get . . ."): Millikan to Greta Millikan, April 18, 1918, RAM, Box 49.8.

Page 84 ("Our relationship to the Government"): Hale to Evelina Hale, March 22, 1918, GEH, Box 81.

Page 84 ("to pay its own expenses . . ."): W. R. Whitney to Hale, June 11, 1918, GEH, Box 43.

Page 84 ("Do you think, for instance"): Hale to Whitney, June 19, 1918, ibid.

Page 85 ("the Throop end of the project"): Hale to Scherer, Oct. 16, 1918, JABS, Box 2.6.

Page 85 ("I feel very sure . . ."): Fleming to Hale, Jan. 30, 1919, in JABS, Box. 2.7.

Page 85 ("Millikan and Noyes were there . . ."): Hale to Scherer, Jan. 29, 1919, JABS, Box 2.7.

Page 85 ("You should take active part"): Hale to Scherer, Feb. 3, 1919, ibid.

Page 85 ("Talk this over . . ."): Hale to Scherer, Feb. 5, 1919, ibid.

Page 86 ("a physical breakdown"): Scherer to W. H. Hays, April 2, 1919, JABS, Box 1.12.

Page 86 ("Oddly enough"): Scherer to Hale, March 21, 1919, GEH, Box 36.

Page 86 ("I have in my pocket . . ."): Scherer to Hale, March 28, 1919, ibid.

Page 86 ("It may be a playful habit . . ."): Scherer to Hale, Sept. 29, 1919, JABS, Box 1.12.

Page 86 ("psychic relief"): Ibid.

Page 86 ("The whole question . . ."): Hale to Scherer, Sept. 24, 1919, JABS, Box 2.7.

Page 86 ("to continue to be . . ."): E. C. Barrett to Scherer, March 24, 1919, ECB, Box 2.10.

Page 87 ("The heavy responsibilities . . ."): Scherer to Fleming, March 10, 1920, JABS, 1.13.

Page 87 ("He should make . . ."): E. B. Wilson to Noyes, Sept. 30, 1920, copy in GEH, Box 44.

CHAPTER 5 Millikan and the Rise of Physics

Page 88 Headquote: Robert A. Millikan, "A New Opportunity in Science," *Science* 50 (1919): 297, reprinted in *Throop College Bulletin* 28 (1919): 3–30.

Page 88 ("a complete loss"): Robert A. Millikan, *Autobiography of Robert A. Millikan* (New York, 1950), p. 14.

Page 89 ("Anyone who can . . ."): Ibid.

Page 89 ("the year"): Ibid., p. 17.

Page 89 ("a strange-appearing . . ."): Ibid., p. 18.

Page 90 ("outlined precisely . . ."): Ibid., p. 24.

Page 90 ("I expected a beautiful . . ."): R. A. Millikan to his parents, Aug. 12, 1895, RAM, Box 53.1.

Page 91 ("As far as I . . ."): Millikan to his parents, Nov. 3, 1895, RAM, ibid.

Page 91 ("I have decided . . ."): Millikan to W. R. Harper, Sept. 22, 1896, RAM, Box 43.11; original in the "Presidents' Papers, 1889–1925" Collection, Regenstein Library, University of Chicago.

Page 91 ("An unforgettable seething . . ."): F. B. Jewett to Millikan, Feb. 17, 1948, RAM, Box 72.4.

Page 91 ("the gamble"): Millikan, *Autobiography*, p. 69.

Page 92 ("I knew"): Ibid.

Page 92 ("I have in mind . . ."): Ibid., p. 75.

Page 93 ("interested even him"): Millikan to Greta Millikan, ca. Dec. 1910, RAM, Box 49.1b.

Page 93 ("the example . . ."): Robert A. Millikan, "A Few Notes about Frank Baldwin Jewett," Sept. 1944, RAM, Box 40.25.

Page 93 ("pure experimentalist"): Robert A. Millikan, "Albert Abraham Michelson," dedication of Michelson Laboratory at Inyokern, May 8, 1948, RAM, Box 61.30.

Page 93 ("he was an intense individualist"): Robert A. Millikan, "Albert Abraham Michelson, 1852–1931," *Biographical Memoirs of the National Academy of Sciences* 19 (1938): 123.

Page 93 ("on the place of . . ."): Quoted in Robert H. Kargon, *The Rise of Robert Millikan* (Ithaca, 1982), p. 35.

Page 93 ("to put physics on the map"): Millikan, *Autobiography*, p. 238.

Page 94 ("You'll ruin yourself"): Quoted in "William R. Smythe," interview by Mary Terrall, 1979, p. 34, Institute Archives.

Page 95 ("I went to Chicago . . ."): Interview of William Vermillion Houston, by Gerald Phillips and W. J. King, 1964, p. 8, American Institute of Physics, New York City, copy in Houston Papers, Rice University Archives, Woodson Research Center, Rice University Library.

Page 96 ("my hands were not clever enough"): Quoted in "Paul S. Epstein," interview by Alice Epstein, 1965, p. 70, Institute Archives.

Page 97 ("I found myself"): Millikan to Greta Millikan, April 11, 1921, RAM, Box 50.5.

Page 98 ("stiff graduate courses") and ("I am still hesitating . . ."): Millikan to G. E. Hale, Aug. 6, 1921 and July 16, 1921, GEH, Box 29.

Page 98 ("We can keep Epstein . . ."): Millikan to Hale, Aug. 6, 1921, ibid.

Page 99 ("a good neighbor policy . . ."): Quoted in "Paul Epstein" interview, p. 120.

Page 100 ("The radios . . ."): Letter to the Editor, *Los Angeles Times*, March 6, 1931.

Page 101 ("January 2, 1931 . . ."): Quoted in Gerald Holton, "The American Physics Community and Albert Einstein," *Minerva* 19 (1981): 578.

Page 101 ("further thought . . ."): A. Einstein to Millikan, Aug. 1, 1931, RAM, Box 39.7.

Page 102 ("Fellow Scientists . . ."): R. C. Tolman, after-dinner remarks, Feb. 5, 1931, RCT, Box 8.13.

Page 103 ("There was no outbreak . . ."): A. A. Noyes to Hale, Oct. 9, 1931, GEH, Box 32.

Page 104 ("If Einstein . . ."): E. C. Barrett to Millikan, Oct. 1, 1931, RAM, Box 27.11.

Page 104 ("Most esteemed Mr. Fleming . . ."): Quoted in Einstein to Millikan, Oct. 19, 1931, ibid.

Page 105 ("Dear Professor Millikan . . ."): Elsa Einstein to Millikan, Nov. 14, 1931, ibid.

Page 105 ("By the authority . . ."): Millikan to Einstein, Dec. 19, 1931, quoted in Executive Council Minutes, in GEH, Box 139.

Page 105 ("Oh, you mean Millikan's school!"): Quoted in "Carl D. Anderson," interview by Harriet Lyle, 1979, p. 15, Institute Archives.

Page 106 ("I do not believe at all . . ."): E. B. Wilson to Hale, Oct. 7, 1919, GEH, Box 44.

Page 106 ("The real purpose . . ."): Quoted in "H. Victor Neher," interview by Rachel Prud'homme, 1982, p. 5, Institute Archives.

Page 107 ("In fact . . ."): Interview of Ira S. Bowen, by Charles Weiner, 1968, p. 7, AIP.

Page 108 ("This productivity"): "Research Activities at CIT," Oct. 6, 1928, p. 3, GEH, Box 6.

Page 108 ("If a man . . ."): Robert A. Millikan, Throop assembly address, Jan. 6, 1920, RAM, Box 20.9.

CHAPTER 6   Biological Work

Page 109 Headquote: Francis C. Sumner to Robert M. Yerkes, April 14, 1923, Robert M. Yerkes Papers, Manuscripts and Archives, Yale University Library.

Page 110 ("In the bonanza years . . ."): Carey McWilliams, *Southern California Country* (New York, 1946), p. 135.

Page 110 ("refugees from America"): Quoted ibid., p. 269.

Page 110 ("We have"): Sumner to Yerkes, April 14, 1923, Yerkes Papers.

Page 111 ("These graduate departments . . ."): H. J. Thorkelson, "Memorandum re California Institute of Technology, Pasadena, California," March 24, 1927, p. 4, GEB, Series 1, Box 612, Folder 6476.

Page 111 ("Aside from its importance . . ."): Arthur Amos Noyes, "Plans for the Development of Biology at the California Institute of Technology," March 20, 1924, RAM, Box 18.7.

Page 111 ("It would probably make . . ."): A. A. Noyes to G. E. Hale, Dec. 22, 1922, GEH, Box 32.

Page 112 ("twice as active a . . ."): Noyes to Hale, Feb. 18, 1923, ibid.

Page 112 ("if anything is done . . ."): R. A. Millikan to Hale, Aug. 28, 1923, GEH, Box 29.

Page 112 ("Your imagination can work . . ."): Ibid.

Page 113 ("to back them up . . ."): "Excerpt from Record of Dr. Rose's Interviews," Oct. 2, 1923, GEB, Series 1, Box 611, Folder 6467, quotes from p. 1.

Page 113 ("The sound thing to do"): Ibid., p. 2.

Page 114 ("attempts to develop an . . ."): Noyes to F. P. Keppel, Oct. 16, 1924, RAM, Box 27.9.

Page 115 ("a high grade of insulin . . ."): Noyes to J. J. Abel, May 31, 1924, John J. Abel Papers, Alan Mason Chesney Medical Archives, Johns Hopkins Medical Institutions, Baltimore.

Page 115 ("Will attack insulin"): Quoted in Noyes to Hale, July 17, 1924, GEH, Box 32.

Page 115 ("is weak on the . . ."): Noyes to Abel, Aug. 6, 1924, Abel Papers, Chesney Medical Archives, Johns Hopkins.

Page 116 ("leaving the question of . . ."): Thorkelson to Millikan, Jan. 29, 1925, GEB, Series 1, Box 612, Folder 6476.

Page 116 ("would talk of nothing . . ."): Noyes to Hale, Feb. 2, 1925, GEH, Box 32.

Page 116 ("The community and country . . ."): Noyes, "Statement," n.d., RAM, Box 18.7, p. 6.

Page 116 ("I am hoping this . . ."): Noyes to Hale, Feb. 2, 1925, GEH, Box 32.

Page 117 ("the risk of losing . . ."): "Notes by Millikan on Policy of C.I.T.," March 14, 1927, p. 5, GEH, Box 29.

Page 118 ("the world's center for . . ."): Thorkelson, "Memorandum," p. 4.

Page 118 ("to build up from . . ."): Millikan, "Notes," p. 4.

Page 119 ("an outstanding man in . . ."): Thorkelson, "Memorandum," p. 6.

Page 119 ("He said *definitely* that . . ."): Noyes to Hale, April 27, 1927, GEH, Box 32.

Page 119 ("I can't tell you . . .") and ("The participation of a . . ."): Hale to T.H. Morgan, May 4, 1927, THM, Box 1, and Morgan to Hale, May 9, 1927, GEH, Box 29.

Page 119 ("greatly interested in our . . ."): Hale to W. Rose, April 24, 1927, GEB, Series 1, Box 612, Folder 6476.

Page 120 ("the needs of the . . ."): Statement by Millikan in margin of letter, Thorkelson to Millikan, April 25, 1927, RAM, Box 25.15.

Page 120 ("suitable men"): Attachment, Morgan to A. Fleming, Aug. 1, 1927, RAM, Box 18.9.

Page 121 ("without doubt the foremost . . ."): Millikan to A. C. Balch, July 5, 1927, RAM, Box 27.9.

Page 122 ("but nothing more"): Millikan to Rose, Oct. 15, 1927, RAM, Box 27.9.

Page 122 ("It seems incredible from . . ."): Morgan to Hale, Oct. 22, 1927, THM, Box 1.

Page 122 ("It may be a . . ."): Hale to Morgan, Oct. 18, 1927, ibid.

Page 123 ("the interest of the Board . . ."): Thorkelson, "Interviews: Mr. A. C. Balch," Nov. 7, 1927, GEB, Series 1, Box 612, Folder 6476.

Page 123 ("I have spoken to . . ."): Millikan to Morgan, Feb. 25, 1928, RAM, Box 18.10.

Page 123 ("not wish to have . . ."): Millikan to Rose, Oct. 15, 1927, RAM, Box 27.9.

CHAPTER 7 Earthquakes

Page 125 Headquote: Charles Richter, "An Instrumental Earthquake Magnitude Scale," *Bulletin of the Seismological Society of America* 25 (Jan. 1935): 1.

Page 128 ("a deep coat . . ."): H. Wood to A. Day, July 1, 1921, HOW, Box 12.7.

Page 128 ("The Earthquake Problem . . ."): Harry O. Wood, "The Earthquake Problem in the Western United States," BSSA 6 (1916): 197–217.

Page 129 ("the instrumental study . . ."): Wood to J. C. Merriam, Sept. 14, 1921, HOW, Box 13.9.

Page 129 ("deduce . . . the places . . ."): Wood, "The Earthquake Problem," p. 208.

Page 130 ("should be telling . . ."): Ibid., p. 209.
Page 132 ("I suppose in general . . ."), ("He knows California . . ."), and ("boosterism"): Day to Wood, June 7, 1921; Wood to Day, June 1, 1921, all in HOW, Box 12.7.
Page 132 ("A less brazen people . . ."): Wood, "Seismology in the United States . . . ," draft, n.d., HOW, Box 20.12.
Page 134 ("Anderson has contrived . . ."): Wood to Day, Nov. 7, 1922, HOW, Box 12.6.
Page 134 ("There ain't no other . . ."): J. Anderson to Wood, Dec. 26, 1924, HOW, Box 11.2.
Page 135 ("The [Carnegie] Institution is committed . . ."): Day to R. A. Millikan, Dec. 6, 1922, RAM, Box 33.1.
Page 135 ("Dr. Millikan has too many . . ."): Wood to Day, Nov. 7, 1922, HOW, Box 12.6.
Page 136 ("The seismogram is indeed . . ."): B. Gutenberg to Wood, April 11, 1925, HOW, Box 3.26.
Page 136 ("permit the [seismology] work . . ."): G. E. Hale to Merriam, June 8, 1924, copy in RAM, Box 33.2.
Page 137 ("We must keep . . ."): Wood to Day, July 26, 1924, HOW, Box 12.4.
Page 138 ("The most serious . . ."): Day to Wood, July 5, 1929, HOW, Box 11.11.
Page 139 ("I think it should . . ."): Wood to Day, Jan. 23, 1926, HOW, Box 12.3.
Page 139 ("as minor side-heads . . ."): Millikan to Merriam, Mar. 16, 1926, RAM, Box 33.3.
Page 139 ("Pasadena Seismological Laboratory"): J. P. Buwalda to Millikan, Sept. 6, 1929, RAM, Box 33.5.
Page 139 ("I think it is . . ."): Day to Wood, July 5, 1929, HOW, Box 11.11.
Page 140 ("brush up his English"): Quoted in Wood to Day, July 27, 1929, ibid.
Page 140 ("on the carpet") and ("We need Gutenberg . . ."): Day to Wood, June 10, 1929, and Wood to Day, Dec. 11, 1929, ibid.
Page 140 ("Not particularly social") and ("These were almost . . ."): Quoted in "Charles F. Richter," interview by Ann Scheid, 1979, pp. 12–14, Institute Archives.
Page 141 ("an ordinary routine worker"): Wood to Day, June 4, 1928, HOW, Box 11.12.
Page 141 ("I would prefer . . ."): Day to Wood, Dec. 3, 1929, HOW, Box 11.11.
Page 142 ("after a Laboratory course . . ."): Beno Gutenberg, "Sketch of Biographical Data," 1945, BG, Box 19.6.
Page 142 ("I considered this paper . . ."): Beno Gutenberg, "Biographical Data," n.d., BG, Box 19.5.
Page 143 ("the travel times . . ."): Beno Gutenberg, "Seismology," in Geological Society of America, Chapter in Geology, 1888–1939: Fiftieth Anniversary Volume (New York, 1941), p. 461.
Page 143 ("I know my boy's . . ."): Quoted in "Herta Gutenberg," interview by Mary Terrall, 1981, p. 5, Institute Archives.
Page 144 ("He was always working"): Ibid., p. 7.
Page 144 ("Now relationships, previously invisible . . ."): E. A. Ansel to Gutenberg, Aug. 2, 1928, BG, Box 1.1.
Page 144 ("I think you will . . ."): Millikan to Gutenberg, May 20, 1930, RAM, Box 33.6.
Page 145 ("The earthquake business . . ."): Quoted in Wood to Day, May 4, 1931, HOW, Box 11.9.
Page 145 ("This is another . . ."): P. Byerly to Gutenberg, Feb. 28, 1931, copy in HOW, Box 14.5.
Page 146 ("We needed something . . ."): Quoted in "Charles F. Richter," p. 33.
Page 147 ("noticed by persons . . ."): Wood and F. Neumann, "Modified Mercalli Intensity Scale of 1931," BSSA 21 (1931): 277–283, on 280.
Page 147 ("In a region such . . ."): Richter, "Earthquake Magnitude Scale," p. 1.
Page 147 ("to make a rough comparison . . ."): Ibid., p. 2.
Page 147 ("Then I got hold . . ."): Quoted in "Charles F. Richter," p. 34.
Page 148 ("Try plotting them on . . ."): Ibid., p. 35.
Page 148 ("If there was anything . . ."): Ibid.
Page 148 ("the most reliable . . ."): Richter, "Earthquake Magnitude Scale," pp. 7–8.
Page 149 ("likely to be felt"): Ibid., p. 28.
Page 150 ("For a shock of . . ."): Ibid., p. 30.
Page 150 ("He would begin by . . ."): Richter, "Memorial," p. 94.

Page 151 ("The whole group of . . ."): Quoted in Wood's progress report to Merriam, 1935–1936, HOW, Box 13.6.
Page 151 ("to withdraw gradually"): Buwalda to Millikan, Sept. 7, 1934, RAM, Box 33.8.
Page 151 ("If we had merely . . ."): Buwalda to Millikan, May 4, 1936, RAM, Box 33.10.
Page 152 ("Richter would have worked . . ."): Wood to Buwalda, Sept. 22, 1938, HOW, Box 1.23.
Page 152 ("As you know . . ."): Buwalda to V. Bush, Sept. 11, 1941, RAM, Box 33.12.

CHAPTER 8   Aeronautics and the Airplane Industry

Page 153 Headquote: Theodore von Kármán, *Aerodynamics: Selected Topics in the Light of Their Historical Development* (Ithaca, 1954), p. 1.
Page 154 ("birdcage with all the . . ."): Ernest E. Sechler, "Structural Developments from Kitty Hawk to the Gossamer Condor," in F. E. C. Culick, ed., *Guggenheim Aeronautical Laboratory at the California Institute of Technology* (San Francisco, 1983), p. 46.
Page 154 ("They did it!"): Quoted in Roger E. Bilstein, *Flight in America, 1900–1983* (Baltimore, 1984), p. 12.
Page 154 ("the epoch-making invention . . ."): Quoted in William F. Durand, "Orville Wright, 1871–1948," *Biographical Memoirs of the National Academy of Sciences* 25 (1949): 262.
Page 155 ("The orders were placed . . ."): Bilstein, *Flight*, p. 36.
Page 157 ("a mathematical physicist to . . ."): H. Bateman to G. E. Hale, April 25, 1916, GEH, Box 4.
Page 157 ("welcome an appointment which . . ."): Ibid.
Page 157 ("in addition to keeping . . ."): R. A. Millikan to Hale, Feb. 18, 1917, GEH, Box 29.
Page 158 ("great deficiency of brains . . ."): E. B. Wilson to Hale, Jan. 22, 1917, GEH, Box 43.
Page 158 ("If you could hire . . ."): Wilson to Hale, Feb. 26, 1917, ibid.
Page 158 ("Position will be Professor . . ."): J. A. B. Scherer to Bateman, April 9, 1917, JABS, Box 1.10.
Page 158 ("I think I should . . ."): Bateman to R. A. Millikan, April 17, 1917, JABS, Box 1.18.
Page 159 ("I doubt the wisdom . . ."): R. A. Millikan to Scherer, April 17, 1917, ibid.
Page 159 ("important research"): Hale to T. Ford, Feb. 10, 1917, GEH, Box 16.
Page 159 ("advice"): "Research in Aeronautics," *Throop College Bulletin* 28 (1919): 104.
Page 159 ("We must beware of . . ."): Quoted in Scherer to Hale, Oct. 27, 1916, JABS, Box 1.9.
Page 160 ("devote his entire time . . ."): Institute Publicity Releases, "Harry Bateman," July 6, 1917, Institute Archives, Historical Files, Box 44.
Page 160 ("relate to the stability . . ."): News Clipping, June 20, 1917, ibid., Scrapbook Collection.
Page 160 ("with a physical laboratory . . ."): Bateman to F. Morley, Feb. 27, 1912, Christopher Darlington Morley Papers, Harry Ransom Humanities Research Center, University of Texas at Austin.
Page 160 ("Bateman, of course, is . . ."): E. C. Barrett to Scherer, Nov. 17, 1917, ECB, Box 2.8.
Page 161 ("I can take seniors . . ."): A. A. Merrill To R. A. Millikan, Sept. 24, 1921, RAM, Box 29.29.
Page 161 ("Spread yourself; it pays . . ."): E. T. Bell to Bateman, July 4, 1927, HB Papers, Box 1.1.
Page 162 ("for advancing the science . . ."): R. A. Millikan to H. L. Guggenheim, Dec. 24, 1925, RAM, 16.5.
Page 162 ("research"): Quoted in Millikan to F. Jewett, Jan. 18, 1925, ibid., Box 16.6.
Page 162 ("there is a very . . ."): Millikan to Guggenheim, Jan. 29, 1926, ibid.
Page 162 ("most interesting"): Guggenheim to Millikan, March 4, 1926, ibid.
Page 162 ("condensed reformulation") and ("an appropriation of funds . . ."): Millikan to Guggenheim, May 14, 1926, and Guggenheim to Millikan, June 7, 1926, all ibid.

Page 163 ("The facilities of the . . ."): *Bulletin of the California Institute of Technology* 35 (1926): 4–5.

Page 163 ("well-trained foreign brain"): P. Epstein to T. von Kármán, July 5, 1926, TVK, Box 8.21.

Page 164 ("to fight it out . . . "): M. Born to von Kármán, Nov. 7, 1922, TVK, Box 3.27.

Page 165 ("had just announced the . . ."): Theodore von Kármán with Lee Edson, *The Wind and Beyond* (Boston, 1967), p. 125.

Page 165 ("I didn't want to . . ."): Quoted in "Arthur L. Klein," interview by Harriett Lyle and John L. Greenberg, 1979–1982, Institute Archives, p. 6.

Page 165 ("They were hopeless, mechanically"): Ibid., p. 9.

Page 165 ("Everybody in the aeronautics . . ."): A. L. Klein, "Closing Remarks," in Culick, ed., *Guggenheim*, p. 85.

Page 166 ("Don't yell!"): A. L. Klein to E. P. Leslie, Jan. 15, 1929, CBM, Box 14.14.

Page 166 ("In fact it was . . ."): Von Kármán, "I come to the United States," autobiographical draft, n.d., pp. 17–18, copy in FJM, Box 17.3.

Page 167 ("The pilot can take . . ."): Clark Millikan to Robert Millikan, June 19, 1926, RAM, Box 16.6.

Page 167 ("to control flight by . . ."): Von Kármán, *The Wind*, p. 124.

Page 167 ("Rules of procedure which . . ."): R. Millikan, "Memorandum," Nov. 1, 1926, RAM, Box 16.6.

Page 168 ("ground looped and broke . . ."): Quoted in "Arthur L. Klein," p. 51.

Page 168 ("Of course any ship . . ."): Merrill to C. Millikan, Nov. 22, 1928, copy in RAM, Box 29.29.

Page 168 ("an Elementary Class") and ("This was supposed to . . ."): Letter from A. E. Raymond, May 15, 1982.

Page 169 ("15 legal-size pages"): Arthur E. Raymond, "Transport Development: The DCs," in Culick, ed., *Guggenheim*, p. 26.

Page 170 ("A sharp corner where . . ."): Von Kármán, *The Wind*, pp. 169–70.

Page 171 ("One of the prominent . . ."): Von Kármán, *Aerodynamics*, pp. 151–52.

Page 171 ("I enjoyed climbing into . . ."): Von Kármán, *The Wind*, p. 170.

Page 171 ("The model is suspended . . ."): "GALCIT," in *Bulletin of the California Institute of Technology* 49 (May 1940): 16.

Page 172 ("of the basic structural . . ."): Raymond, "Transport Development," p. 24.

Page 172 ("I think what they . . ."): Quoted in "Arthur L. Klein," p. 20.

Page 173 ("secret"): A. L. Klein to C. N. Monteith, Sept. 26, 1931, TVK, Box 57.16.

Page 173 ("school vacation [when] it . . ."): Klein to R. J. Minshall, April 26, 1932, ibid.

Page 173 ("malicious reports"): Monteith to Raymond, Aug. 30, 1934, copy in TVK, Box 57.17.

Page 174 ("the necessity for keeping . . ."): Monteith to Klein, Oct. 24, 1934, ibid.

Page 174 ("I discussed with you . . ."): C. B. Millikan to D. Douglas, Sept. 19, 1936, CBM, Box 10.9.

Page 175 ("the constant contact of . . ."): Clark B. Millikan, "On the Results of Aerodynamic Research and their Application to Aircraft Construction" (Paper presented at Lilienthal Society in Berlin, 1936), pp. 36–37, ibid.

Page 176 ("that for the time . . ."): "GALCIT," p. 16.

Page 176 ("Personally, . . ."): Von Kármán to Col. H. H. Zornig, June 13, 1939, TVK, Box 91.14.

Page 176 ("As you are probably . . ."): Von Kármán to A. Butt, July 28, 1939, TVK, Box 69.25.

Page 177 ("It is hard to . . ."): Raymond to von Kármán, May 15, 1940, TVK, Box 59.4.

Page 177 ("aeronautical engineering was in . . ."): Quoted in "Arthur L. Klein," p. 17.

CHAPTER 9 Atoms, Molecules, and Linus Pauling

Page 178 Headquote: "Interviews-Dr. Linus Pauling," Sept. 20, 1946, p. 2, RAC, Record Group 1.1, Series 205, Box 7, Folder 97.

Page 178 ("the fuzziness of chemistry . . ."): Linus Pauling, "Fifty Years of Progress in Structural Chemistry and Molecular Biology," *Daedalus* (Fall 1970): 989.

Page 179 ("atoms were assigned a . . ."): Ibid., p. 990.

Page 179 ("to understand the physical . . ."): Ibid., p. 988.

Page 180 ("quantitative information about the . . ."): Linus Pauling, "Roscoe Gilkey Dickinson," *Science* 102 (1945): 216.

Page 181 ("After two months work . . ."): J. D. H. Donnay, "Presentation of the 1967 Roebling Medal of the Mineralogical Society of America to Linus Pauling," *American Mineralogist* 53 (March–April 1968): 527.

Page 181 ("one of our Fellows . . ."): A. A. Noyes to G. E. Hale, Feb. 18, 1923, GEH, Box 32.

Page 182 ("The faculty seems to . . ."): Linus Pauling to F. Allen, Oct. 25, 1924, reproduced in Derek Davenport, "Vintage Pauling," *Journal of Chemical Education* 59 (Dec. 1982): 716.

Page 182 ("I don't know . . ."): Pauling, "Fifty Years of Progress," p. 992.

Page 185 ("I felt that the . . ."): Linus Pauling, "Fifty Years of Physical Chemistry in the California Institute of Technology," *Annual Review of Physical Chemistry* 16 (1965): 9.

Page 185 ("Is quantum mechanics . . ."): Quoted in Linus Pauling," interview by John L. Heilbron, March 27, 1964, session two, p. 13, Archives for the History of Quantum Physics, Office for History of Science and Technology, University of California, Berkeley.

Page 186 ("I thought there was . . ."): Ibid., p. 17.

Page 186 ("Some people"): Pauling to A. A. Noyes, Dec. 17, 1926, quoted ibid., pp. 31–32.

Page 187 ("Well, there was this . . ."): Ibid., p. 15.

Page 188 ("Modern chemistry insensibly merges . . ."): "Seat of Science," *Fortune*, July 1932, p. 22.

Page 188 ("Pauling was useful to . . ."): Quoted in "Joint Interview with James Bonner, Sterling Emerson, Norman Horowitz, Donald Poulson," interview by Judith Goodstein, Harriet Lyle, and Mary Terrall, Nov. 6, 1978, p. 35, Institute Archives.

Page 189 ("This knowledge . . ."): Linus Pauling, "A Program of Research in Structural Chemistry," 1932, RAC, RG 1.1, Series 205, Box 5, Folder 70.

Page 189 ("I had initiated . . ."): Pauling, "Fifty Years of Progress," p. 1002.

Page 190 ("All my life"): "Interviews-Dr. Linus Pauling," Sept. 20, 1946, p. 2, RAC, RG 1.1, Series 205, Box 7, Folder 97.

Page 190 ("I try to identify . . ."): "Interviews-Dr. Linus Pauling," July 2, 1951, p. 1, ibid.

Page 191 ("We took these distances . . ."): Ibid., p. 2.

Page 191 ("He knew his atoms . . ."): J. D. Bernal, "The Pattern of Linus Pauling's Work in Relation to Molecular Biology," in Alexander Rich and Norman Davidson, eds., *Structural Chemistry and Molecular Biology* (San Francisco, 1968), p. 378.

Page 192 ("his helical hypothesis was . . ."): Ibid.

Page 192 ("Were all the rest . . ."): Warren Weaver's diary, Oct. 23–25, 1933, p. 79, RG 12.1, RFA.

Page 192 ("although we can weigh . . ."): Linus Pauling, "Molecular Architecture and Medical Progress," in Warren Weaver, ed., *The Scientists Speak* (New York, 1947), p. 113.

CHAPTER 10   The Thomas Hunt Morgan Era in Biology

Page 193 Headquote: Thomas Hunt Morgan, "The New Division of Biology," *Bulletin of the California Institute of Technology* 36 (1927): 86–87.

Page 193 ("What in hell are . . ."): Quoted in "Henry Borsook," interview by Mary Terrall, 1981, p. 5, Institute Archives.

Page 194 ("just next to God'") and ("close to Heaven"): Theodosius Dobzhansky, "Reminiscences," Part I, annotated transcript of an interview by Barbara Land, 1962, p. 221, Columbia University Oral History Project, housed at the American Philosophical Society Library, Philadelphia.

Page 194 ("a very small, poorly . . ."): Dobzhansky, "Reminiscences," p. 239.

Page 195 ("fits and spurts"): Ibid., p. 260.

Page 195 ("In fact, one of . . ."): Thomas Hunt Morgan, "Calvin Blackman Bridges, 1889–1938," *Biographical Memoirs of the National Academy of Sciences* 22 (1941): 40.

Page 195 ("immediately announced"): Ibid., p. 33.

Page 195 ("had the best 'eye' . . ."): Alfred Henry Sturtevant, *A History of Genetics* (New York, 1965), p. 51.

Page 196 ("for I could see . . ."): Alfred Henry Sturtevant, "Biographical Notes," ca. 1962, Anne Roe Collection, American Philosophical Society Library.

Page 196 ("So this one room . . ."): Dobzhansky, "Reminiscences," pp. 239–40.

Page 197 ("Everybody did his own . . ."): Alfred Henry Sturtevant, "Personal Recollections of Thomas Hunt Morgan," unpublished talk, April 25, 1966, AHS, Box 3.22, pp. 4–5.

Page 197 ("There are a lot . . ."): Quoted in Elof Axel Carlson, *Genes, Radiation, and Society: The Life and Work of H. J. Muller* (Ithaca, 1981), p. 63.

Page 197 ("to suggest a topic."): Dobzhansky, "Reminiscences," pp. 243–44.

Page 199 ("elements"): V. Kruta and V. Orel, "Johann Gregor Mendel," in *Dictionary of Scientific Biography*, 16 vols. (New York, 1970–80), 9:277-83.

Page 199 ("Whereas it was difficult . . ."): John A. Moore, "Thomas Hunt Morgan—The Geneticist," *American Zoologist* 23 (1983): 862–63.

Page 200 ("Muller was a very . . ."): Quoted in Anne Roe Collection, Notes on Rorschach interpretation, ZG2, ca. 1962, American Philosophical Society Library.

Page 200 ("I suddenly realized . . ."): Alfred Henry Sturtevant, *History of Genetics*, p. 47.

Page 201 ("first job on *Drosophila*"): Quoted in "A. H. Sturtevant," transcript of an interview by Anne Roe, Dec. 1962, ZG 2-2, American Philosophical Society Library.

Page 201 ("For Morgan himself the . . ."): Curt Stern, "The Continuity of Genetics," *Daedalus* 70 (Fall 1970): 899.

Page 202 ("I am writing to . . ."): T. H. Morgan to G. E. Hale, Aug. 15, 1927, copy in RAM, Box 18.9.

Page 202 ("Only through an exact . . .") and ("The best chance"): Thomas Hunt Morgan, "The Relation of Biology to Physics," *Science* 65 (March 4, 1927): 214, 217.

Page 203 ("with distinct literary ability"): Morgan to R. A. Millikan, May 4, 1928, RAM, Box 18.10.

Page 204 ("three brilliant young men"): Millikan to Morgan, May 16, 1928, ibid.

Page 204 ("I . . . consider the matter"): Morgan to Millikan, May 28, 1928, ibid.

Page 204 ("collected about himself . . ."): Ibid.

Page 204 ("it was shocking"): Eva M. Jacoby to Susan Trauger, Oct. 4, 1979, Institute Archives.

Page 205 ("a man of wide . . ."): Dobzhansky, "Reminiscences," p. 249.

Page 205 ("The question of Morgan's . . ."): Letter from Norman Horowitz, June 21, 1990.

Page 205 ("But time and again . . ."): Dobzhansky, "Reminiscences," p. 254.

Page 205 ("But he was never mean"): Sturtevant, "Personal Recollections," p. 6.

Page 205 ("to look over the ground . . ."): Morgan to Warren Weaver, Jan. 11, 1934, RFA, Record Group 1.1, Series 205, Box 5, Folder 72.

Page 206 ("He has announced to . . ."): "Excerpt from W. E. Tisdale's Log," May 9, 1934, ibid.; see also, Tisdale to Weaver, May 6, 1936; ibid., Series 205D, Box 6, Folder 75.

Page 206 ("gave somewhat the impression . . ."): Harry M. Miller's diary, June 7, 1934, RG 12.1, RFA.

Page 206 ("the sacred flame"): Dobzhansky, "Reminiscences," p. 289.

Page 206 ("thought she had . . ."): Quoted in Joint Interview with James Bonner, Sterling Emerson, Norman Horowitz, Donald Poulson, interview by Judith Goodstein, Harriet Lyle, and Mary Terrall, Nov. 6, 1978, p. 3, Institute Archives.

Page 207 ("was connected by a . . ."): Quoted in "James F. Bonner," interview by Graham Berry, 1982, p. 11, Institute Archives.

Page 207 ("The students sat in . . ."): Ibid., p. 10.

Page 207 ("Just what I expected . . ."): Quoted in Dobzhansky, "Reminiscences," pp. 196–97.

Page 208 ("even if there were . . ."): Thomas Hunt Morgan, "The Relation of Genetics to Physiology and Medicine," in *Les Prix Nobel en 1933* (Stockholm, 1935), p. 3, copy in THM, Box 2.

Page 208 ("asked the question, 'Dobzhansky' . . ."): Quoted in Dobzhansky, "Reminiscences," p. 299.

Page 209   ("giant chromosomes"): Letter from Edward C. Lewis, June 21, 1990.
Page 209   ("Dobzhansky, show me the . . ."): Quoted in Dobzhansky, "Reminiscences," p. 328.
Page 209   ("Bridges's map . . ."): Letter from E. C. Lewis, June 21, 1990.
Page 209   ("in a sense competing"): Theophilus S. Painter to D. F. Jones, Oct. 15, 1934, THM, Box 1.
Page 210   ("I do not intend . . ."): Ibid.
Page 210   ("We are not interested . . ."): Morgan to Jones, Oct. 31, 1934, ibid.
Page 210   ("outburst"): Morgan to Jones, Nov. 1, 1934, ibid.
Page 211   ("it would come out . . ."): Quoted in "Joint Interview," p. 18.
Page 211   ("I don't know whether . . ."): Quoted ibid.; Calvin B. Bridges, "The Bar 'Gene' a Duplication," Science 83 (Feb. 28, 1936): 210–11.
Page 211   ("the attention of American . . ."): Hermann J. Muller, "Bar Duplication," Science 83 (May 29, 1936): 528.
Page 212   ("said that as long . . ."): James Bonner, quoted in "Joint Interview," p. 31.
Page 212   ("was that genetics was . . ."): Ibid.
Page 212   ("Morgan's objective in biology . . ."): Sturtevant, "Personal Recollections," p. 7.
Page 212   ("Willy Fowler, who will . . ."): Quoted in "Joint Interview," pp. 9–10.

CHAPTER 11   Astronomy and the 200-Inch Telescope

Page 213   Headquote: Memorandum, Warren Weaver to Raymond B. Fosdick, Dec. 26, 1947, RFA, Record Group 1.2, Series 205D, Box 5, Folder 34.
Page 214   ("Looking ahead and speculating . . ."): Quoted in Helen Wright, Explorer of the Universe: A Biography of George Ellery Hale (New York, 1966), p. 388.
Page 214   ("a large physical laboratory . . ."): Quoted in Helen Wright, "George Ellery Hale," Dictionary of Scientific Biography, 6:28.
Page 215   ("I had not the . . ."): George Ellery Hale, "Biographical Notes," Feb. 8, 1933, p. 24, GEH, Box 92.
Page 215   ("a born adventurer, with . . ."): G. E. Hale to J. C. Merriam, March 29, 1923, GEH, Box 28.
Page 216   ("interesting someone like Yerkes . . ."): Hale to L. F. Hartman, Aug. 5, 1927, GEH, Box 69.
Page 216   ("Like buried treasure, the . . ."): George Ellery Hale, "The Possibilities of Large Telescopes," in Harper's Monthly Magazine 156 (April 1928): 639–40.
Page 217   ("Do you want a . . ."): Quoted in Hale, "Biographical Notes," p. 26.
Page 217   ("He has decided to . . ."): Hale to Evelina Hale, April 12, 1928, GEH, Box 81.
Page 217   ("It was a curious . . ."): Ibid., April 15, 1928.
Page 217   ("close cooperation"): Hale to W. Rose, April 16, 1928, GEH, Box 35.
Page 218   ("They [Rockefeller] do not like . . ."): Hale to Evelina Hale, April 12, 1928, GEH, Box 81.
Page 218   ("all blow up"): Hale to Evelina Hale, ibid.
Page 218   ("drop the matter . . ."): Telegram, A. A. Noyes to Hale, April 26, 1928, GEH, Box 32.
Page 218   ("As I understand Dr. Rose"): Quoted in Helen Wright, Palomar (New York, 1952), p. 55; A. A. Noyes, "Account of the Interviews of Arthur A. Noyes with Dr. John C. Merriam and Others in Regard to the Astrophysical Laboratory of the California Institute," attachment to Noyes's letter to Hale, Sept. 22, 1928, in GEH, Box 32.
Page 219   ("had not been started . . ."): Warren Weaver, "Interviews—Max Mason," Jan. 31, 1937, GEB, Series 1, Subseries 4, Box 612, Folder 6473.
Page 219   ("He was by nature . . ."): Chester Stock, "John Campbell Merriam, 1869–1945," Biographical Memoirs of the National Academy of Sciences 26 (1951): 216.
Page 220   ("I think it is peculiarly . . ."): W. S. Adams to Hale, April 30, 1928, attachment to Rose's letter to Hale, May 17, 1928, GEH, Box 35.
Page 220   ("I will come back . . ."): Hale to Evelina Hale, May 3, 1928, ibid., Box 81.
Page 220   ("If Merriam does not . . ."): Ibid., May 8, 1928.
Page 221   ("Carty gave him . . ."): Ibid., May 11, 1928.

Page 221 ("Dr. Merriam explained that . . ."): Warren Weaver's diary, March 24, 1939, RG 12.1, RFA.

Page 221 ("turned the heat on . . ."): Ibid.

Page 221 ("to get the telescope"): Quoted in Wright, *Explorer*, p. 394.

Page 222 ("Can you come to . . ."): Hale to Porter, Nov. 17, 1928, GEH, Box 33.

Page 222 ("What a privilege to . . ."): Rose to Hale, Nov. 14, 1929, ibid., Box 35.

Page 224 ("many years of experience . . ."): C. S. McDowell, "Final Report on the 200-inch Telescope Project," Dec. 30, 1938, p. 17, GEB, Series 1, Subseries 4, Box 612, Folder 6473, RAC.

Page 225 ("If the present informal . . ."): Fred E. Wright, "The 200-Inch Telescope Situation," Jan. 29, 1934, JAA, Box 1.7.

Page 226 ("This great undertaking of . . ."): Hale to E. Root and H. S. Pritchett, April 2, 1934, GEH, Box 35.

Page 226 ("had felt no compunction . . ."): W. Weaver, "Interviews—Max Mason," Jan. 31, 1937, GEB, Series 1, Subseries 4, Box 612, Folder 6473.

Page 227 ("Oh, some stars"): Quoted in Wright, *Palomar*, p. 169.

CHAPTER 12 Nuclear Reactions

Page 228 Headquote: William A. Fowler, quoted in Jane Dietrich, "William A. Fowler, Nobel Laureate 1983," *Engineering and Science* 47 (Nov. 1983): unpaginated insert.

Page 229 ("too remote for profitable speculation"): George Ellery Hale, *The Study of Stellar Evolution* (Chicago, 1908), p. 213.

Page 229 ("combined to form more . . ."): Quoted in Karl Hufbauer, "Astronomers Take Up the Stellar-Energy Problem, 1917–1920," *Historical Studies in the Physical Sciences* 11 (1981): 301.

Page 230 ("I can tell you"): Quoted in "William A. Fowler," interview by John Greenberg, 1986, p. 88, Institute Archives.

Page 230 ("For those of us . . ."): William A. Fowler, "Phyphty Years of Phun and Physics in Kellogg," *Engineering and Science* 45 (March 1982): 20.

Page 231 ("Lauritsen's qualifications in nuclear . . ."): M. Tuve to J. Slepian, May 7, 1937, Tuve Papers, Library of Congress, Washington, D.C.

Page 231 ("some things in general . . ."): Interview of Charles C. Lauritsen, by Charles Weiner, June 27, 1966, p. 3, American Institute of Physics, New York City.

Page 231 ("I had no intention . . ."): Ibid.

Page 233 ("From Robert Oppenheimer's account . . ."): E. Lawrence to C. C. Lauritsen, Feb. 9, 1933, Carton 10, Folder 36, Ernest Lawrence Papers, Bancroft Library, UC Berkeley.

Page 233 ("I have noticed a . . ."): Lawrence to Tuve, Feb. 18, 1933, Tuve Papers.

Page 234 ("with negative results"): Tuve to Lauritsen, March 29, 1933, CCL, Box 1.8.

Page 234 ("OK, Charlie, lend me . . ."): William A. Fowler, "Nuclear Astrophysics—Today and Yesterday," *Engineering and Science* 33 (June 1969): 8.

Page 234 ("This marked difference in . . ."): Ibid., p. 9.

Page 236 ("When Bethe came out . . ."): Interview of William A. Fowler, by Charles Weiner, June 8 and 9, 1972, p. 39, AIP.

CHAPTER 13 The Rockets' Red Glare

Page 239 Headquotes: F. B. Jewett to R. A. Millikan, Oct. 19, 1942, RAM, Box 6.8; C. C. Lauritsen to R. C. Tolman, June 24, 1940, Tolman Files, Record Group 227, National Archives, Washington, D.C.; Millikan to Jewett, Aug. 6, 1945, RAM, Box 6.9.

Page 241 ("If I had played . . ."): Quoted in Earnest C. Watson, "World War II Rocket Program of the California Institute of Technology," interview by A. B. Christman, Feb. 19 and 20, 1970, p. 3, Naval Weapons Center, China Lake, Calif., copy in Institute Archives.

Page 242 ("a great mixture of . . ."): Quoted in Judith R. Goodstein, "Richard Chace Tolman," in *Dictionary of American Biography*, suppl. 4 (New York, 1974), p. 838.

Page 242 ("I had to set . . ."): Quoted in Watson, "World War II Rocket Program," p. 22.

Page 243 ("as long as the . . ."): Richard C. Tolman, "Memorandum on Items to Be Discussed by Dr. Tolman with Dr. Millikan," June 27, 1940, RAM, Box 21.2.

Page 243   ("I want to know . . ."): Tolman to C. C. Lauritsen, June 21, 1940, Tolman Papers, Box 6, National Archives.

Page 243   ("these good contacts with . . ."): Ibid., July 13, 1940.

Page 244   ("at least possible"): Millikan to C. C. Lauritsen, Sept. 9, 1940, RAM, Box 21.2.

Page 244   ("I am convinced . . ."): Millikan to H. V. Neher, Dec. 2, 1940, RAM, Box 21.2.

Page 245   ("But the upshot was . . ."): William Fowler, "Associates Talk," May 3, 1950, WAF, South Pacific trip notebook.

Page 246   ("I can't for the life . . ."): Quoted in Matthew Josephson, *Infidel in the Temple* (New York, 1967), p. 463.

Page 246   ("great arsenal of democracy"): Quoted in William E. Leuchtenburg, "Franklin D. Roosevelt and the New Deal, 1932–1940," in Arnold A. Offner, ed., *America and the Origins of World War II* (Boston, 1971), p. 94.

Page 247   ("on proximity fuses for . . ."): Interview of William A. Fowler by Charles Weiner, 1972, p. 78, AIP.

Page 247   ("It was claimed in . . ."): Quoted in Thomas Lauritsen, "Wartime Rocket Work of the California Institute of Technology," interview by A. B. Christman, June 6, 1969, p. 6, Naval Weapons Center, copy in Institute Archives.

Page 248   ("Probably the most important . . ."): Ibid.

Page 248   ("There is, as you know . . ."): C. C. Lauritsen to C. B. Millikan, April 18, 1941, CBM, Box 12.11.

Page 249   ("I do not find . . ."): Tolman to V. Bush, April 4, 1941, Tolman Papers, Box 6, National Archives.

Page 249   ("felt that they had . . ."): Charles C. Lauritsen, "Some Facts Concerning Rocket Development in Division A, N.D.R.C.," Jan 30, 1942, pp. 4–5, WAF, Box 15.

Page 250   ("Does the Army and Navy . . ."): Quoted in Albert B. Christman, *Sailors, Scientists, and Rockets* (Washington, D.C., 1971), p. 258.

Page 250   ("as promptly as possible . . ."): Bush to Tolman, May 7, 1941, Tolman Papers, Box 6, National Archives.

Page 250   ("Dr. Bush accepted this . . ."): C. C. Lauritsen, "Some Facts," p. 6.

Page 251   ("could not in any . . ."): Ibid.

Page 251   ("I think Charlie worked . . ."): Quoted in Thomas Lauritsen, "Wartime Rocket Work," p. 8.

Page 252   ("Charlie had always been . . ."): Quoted in William Fowler, "Caltech Rocket Program of World War II and the Naval Ordnance Test Station," interview by A. B. Christman, January 1969, pp. 14–15, Naval Weapons Center, copy in Institute Archives.

Page 252   ("development and testing of . . ."): Minutes of Meeting, National Defense Research Committee, Section H, Division A, Aug. 27, 1941, p. 6, copy in Tolman Papers, Box 16, National Archives.

Page 252   ("approved hook, line, and sinker"): Quoted in Fowler, "Caltech Rocket Program," p. 17.

Page 253   ("I didn't take the leadership . . ."): Quoted in Watson, "World War II Rocket Program," pp. 23–24.

Page 253   ("Earnest was just the opposite . . ."): Quoted in Fowler, "Caltech Rocket Program," p. 17.

Page 254   ("In about two weeks . . ."): C. C. Lauritsen, "Some Facts," p. 7.

Page 254   ("By operating the presses . . ."): Lauritsen to Tolman, Dec. 19, 1941, Tolman Papers, Box 16, National Archives.

Page 255   ("filling powder bags . . ."): Fowler to the author, Nov. 7, 1990.

Page 256   ("Very few people realized . . ."): Quoted in Watson, "World War II Rocket Program," p. 39.

Page 256   ("So the Navy . . ."): Quoted in Christman, *Sailors*, p. 131.

Page 256   ("Mousetrap ammunition"): John E. Burchard, *Rockets, Guns and Targets* (Boston, 1948), p. 99; James P. Baxter, *Scientists against Time* (Boston, 1946), p. 205.

Page 257   ("Charge off more ammunition . . ."): W. Smythe to W. Fowler, July 26, 1942, WAF, new letters file.

Page 258   ("The obvious solution is . . ."): Lauritsen to Tolman, June 20, 1942, copy in WAF, Box 16.

Page 258 ("No, because it doesn't . . ."): Quoted in Burchard, *Rockets*, p. 132.

Page 258 ("The results against exposed . . ."): William Fowler, "Discussion of Rocket Applications in South Pacific . . . ," March, 1944, WAF, Box 19.

Page 259 ("literally cussed a blue streak . . ."): W. Fowler to C. C. Lauritsen, March 8, 1944, WAF, Box 19.

Page 259 ("Coming out of the dive . . ."): Quoted in Burchard, *Rockets*, p. 169.

Page 259 ("We felt that we . . ."): Quoted in "William A. Fowler," p. 80.

Page 259 ("The new shell with . . ."): Quoted in Baxter, *Scientists*, p. 236.

Page 260 ("Right then and there . . ."): Quoted in "William A. Fowler," p. 86.

CHAPTER 14   The DuBridge Era at Caltech

Page 261 Headquotes: Quoted in "Rodman W. Paul," interview by Carol Bugé, 1982, p. 77, Institute Archives; Lee A. DuBridge, "The Responsibility of the Scientist," Caltech commencement address, June 13, 1947, LAD, Box 218; J. R. Page to W. Weaver, Oct. 25, 1961, Historical Files, Box 81.

Page 263 ("looked upon at Caltech . . ."): Frank Malina, personal communication.

Page 264 ("did as well [then] . . ."): Quoted in "Robert F. Bacher," interview by Mary Terrall, 1983, p. 138, Institute Archives.

Page 265 ("definitely and completely out . . ."): J. Page to W. Weaver, Oct. 24, 1945, Historical Files, Box 87.

Page 265 ("had deteriorated mentally"): Page to Weaver, Oct. 25, 1961, ibid., Box 81.

Page 265 ("RAM understands it in . . ."): Page to Weaver, Oct. 24, 1945, ibid., Box 87.

Page 266 ("It seems to me . . ."): Ibid.

Page 266 ("DuBridge is an excellent . . ."): Weaver to Page, Nov. 9, 1945, ibid.

Page 266 ("Mason talked rapidly, wrote . . ."): Lee A. DuBridge, "Memories," personal recollections, 1979, p. 3, copy in Institute Archives.

Page 267 ("sell his Jews to . . ."): Quoted in Victor F. Weisskopf, *The Privilege of Being a Physicist* (New York, 1989), p. 203.

Page 267 ("Bringing Vicky to America. . ."): DuBridge, "Memories," p. 28.

Page 268 ("By the fall of 1938"): Ibid., p. 17.

Page 268 ("You will find it . . ."): I. I. Rabi to L. A. DuBridge, Feb. 26, 1946, LAD, Box 111.3.

Page 269 ("that at the California . . ."): V. Bush to Page, April 15, 1946, Historical Files, Box 87.

Page 269 ("No one knows exactly . . ."): C. Overhage to Page, Sept. 16, 1945, ibid.

Page 269 ("Lee, you've GOT . . ."): DuBridge, "Memories," p. 39.

Page 269 ("He said the offer . . ."): Telephone message, n.d., Historical Files, Box 87.

Page 270 ("decided to consider favorably"): Quoted in Page's telegram to H. S. Mudd, April 16, 1946, ibid.

Page 270 ("What is a Chairman . . ."): Quoted in "Lee A. DuBridge," interview by Judith Goodstein, 1982, p. 15, Institute Archives.

Page 270 ("In any event . . ."): Page to Weaver, Oct. 24, 1945, Historical Files, Box 87.

Page 271 ("[Physics] is close to . . ."): DuBridge to Page, April 29, 1946, ibid.

Page 272 ("reception"): Quoted in "Robert F. Bacher," p. 128.

Page 272 ("had to do with . . ."): Quoted in "Hans A. Bethe," interview by Judith Goodstein, 1982, p. 26, Institute Archives.

Page 273 ("One of the reasons . . ."): Robert Bacher, "A Few Highlights in Teaching and Research . . . ," Aug. 23, 1957, LAD, Box 30.12.

Page 273 ("Feynman was a terrible loss"): Quoted in "Hans A. Bethe," p. 21.

Page 274 ("I never considered thinking . . ."): Quoted in "Robert F. Bacher," p. 148.

Page 275 ("pretty tough character"): Ibid., p. 143.

Page 275 ("He believed in his people . . ."): Quoted in "The Purists," *Time*, May 16, 1955, p. 79.

Page 275 ("I would go to . . ."): Quoted in "Robert P. Sharp," interview by Graham Berry, 1981, p. 43, Institute Archives.

Page 276 ("locally tax-supported institutions . . ."): Robert A. Millikan, "The Past Objectives of the California Institute of Technology," in *Bulletin of the California Institute of Technology* 55 (1946): 21.

Page 276 ("for its support *solely* . . ."): Lee A. DuBridge, "Science and National Security," ibid., 58 (1949): 18.

Page 277 ("not the fault of . . ."): Quoted in Clayton R. Koppes, *JPL and the American Space Program* (New Haven, 1982), p. 48.

INDEX

Abel, John J., 114–15, 116
abolitionists, 26
Adams, Walter, 217, 220, 225, 226
aeronautics:
  as academic discipline, 156
  as amateurs' field, 161–62
  first transcontinental flight and, 155
  Wright brothers and, 154, 155, 156
  see also aeronautics department; airplane
    design; airplane industry
aeronautics department (Caltech), 156–57
  academic vs. industrial careers of graduates
    from, 169
  advanced degrees awarded by, 175
  airplane design classes of, 168–69
  airplane industry's ties to, 173–77
  establishing of, 156–62, 162–63
  first doctorate awarded by, 169
  Guggenheim funding of, 117, 162–63
  Jet Propulsion Laboratory and, 262–63
  von Kármán as head of, 156, 164
  wind tunnels of, 158, 159, 160, 161, 163,
    ·164–66, 170, 171, 172, 173, 174, 175–
    76
Agriculture Department, U.S., 242
aircraft rockets, 245, 255, 259
airplane design:
  all-metal, 172
  cantilevered wings in, 156
  cowlings in, 172
  of DC-1, 169, 170–72
  of DC-3, 172–73
  fillets in, 170–71
  Merrill's course in, 161
  movable vertical tails in, 154
  movable wings in, 161, 166–67
  performance analysis and, 169, 170
  vortex shedding in, 170–71
  wing warping in, 154
  Wright brothers and, 154
  see also aeronautics; aeronautics depart-
    ment; airplane industry
airplane industry:
  Caltech graduates hired by, 169
  Caltech's ties to, 173–74
  DC-3's effect on, 172
  growth of, 153–54, 155, 172
  production figures of, 155, 172
  in World War I, 155

  see also aeronautics; aeronautics depart-
    ment; airplane design
Alles, Gordon A., 114
alpha helix, 191–92
Altenburg, Edgar 197
American Chemical Society, 188
American Naturalist, 210
amino acids, 190–91
Anderson, Carl, 105, 106–7, 252, 271
Anderson, Ernest, 203, 204, 206, 207
Anderson, John C., 131, 133, 134, 141, 221–22,
  225, 227
antiaircraft rockets, 251
antielectrons (positrons), 105, 107
anti-Semitism, 98, 127, 144, 164, 204–6
antisubmarine rockets, 255–56
armor-piercing bombs, 249
Arms and Mudd Laboratories of the Geolog-
  ical Sciences, 240
Army, U.S.:
  Air Corps, 263
  Jet Propulsion Laboratory and, 263–64,
    277
  Ordance Department, 263–64
  Signal Corps, 154, 159
Arnold, Henry H., 263
Arnold, Ralph, 131–32, 133
Arnold, Weld, 262
artificial parthenogenesis, 67
astronomy,
  at Caltech, 274–75
  see also Palomar Observatory telescope
Astronomy and Physics Club, 99
Astrophysical Journal, 44
Astrophysical Laboratory, 222, 240, 255
Atombau und Spektrallinien (Sommerfeld),
  266
atomic bomb project, 260, 272, 273, 274
Atomic Energy Commission, 272
atomic theory:
  in ancient Greece, 183
  chemists' vs. physicists' approach to, 184
  Rutherford-Bohr, 97, 229
  see also chemical-bond theories

Bacher, Robert, 262, 270, 278
  background of, 272
  on Christy, 274

303